# LABWORK TO LEADERSHIP

# LABWORK TO LEADERSHIP

A Concise Guide to Thriving in the Science Job You Weren't Trained For

JEN HEEMSTRA

**HARVARD UNIVERSITY PRESS**
CAMBRIDGE, MASSACHUSETTS & LONDON, ENGLAND
2025

Copyright © 2025 by the President and Fellows of Harvard College
All rights reserved
Printed in the United States of America

Second printing

*Library of Congress Cataloging-in-Publication Data*

Names: Heemstra, Jen, author.
Title: Labwork to leadership : a concise guide to thriving in the science
   job you weren't trained for / Jen Heemstra.
Description: Cambridge, Massachusetts : London, England : Harvard
   University Press, 2025. | Includes bibliographical references and index.
   | Identifiers: LCCN 2024037412 | ISBN 9780674258631 (cloth) | 9780674300705
   (epub) | 9780674300699 (pdf)
Subjects: LCSH: Leadership. | Management. | Laboratories. | Career
   development. | Mentoring in science.
Classification: LCC HD57.7 .H3965 2025 | DDC 658.4/092--dc23/eng/20240925
LC record available at https://lccn.loc.gov/2024037412

EU GPSR Authorised Representative
LOGOS EUROPE, 9 rue Nicolas Poussin, 17000, LA ROCHELLE, France
E-mail: Contact@logoseurope.eu

*To the early-career researchers
who are leading us into
a better future*

# CONTENTS

**INTRODUCTION**   1

## I. SELF-LEADERSHIP

1   **TIME MANAGEMENT**   11
Doing What's Important and Keeping It Organized

2   **LEADERSHIP STRENGTHS AND STYLES**   29
What Makes You Weird?

3   **GOALS**   47
Where Do *You* Want to Go?

4   **MOTIVATION**   61
What to Do When You Don't Want to Do Anything

5   **RESILIENCE**   76
Coping with Failure and Success

6   **RECEIVING FEEDBACK**   91
We All Need Some Coaching to Be Our Best

## II. LEADING OTHERS

| | | |
|---|---|---|
| 7 | **OWNERSHIP**<br>Because Leading Should Be a Team Sport | 113 |
| 8 | **ENVIRONMENT**<br>Creating a Place Where Everyone Can Thrive | 127 |
| 9 | **GIVING FEEDBACK**<br>Candor Doesn't Have to Be Unkind | 143 |
| 10 | **CONFLICT RESOLUTION**<br>What Do You *Really* Want? | 159 |
| 11 | **ETHICAL LEADERSHIP**<br>Playing by the Rules | 180 |
| 12 | **COMMUNICATION**<br>You Said It but Did They Hear It? | 197 |

## III. COACHING FUTURE LEADERS

| | | |
|---|---|---|
| 13 | **PASS IT ON**<br>Cultivating the Next Generation of Leaders | 217 |
| 14 | **VULNERABILITY**<br>Nobody Is Perfect and That's a Good Thing | 229 |
| 15 | **CHALLENGING SITUATIONS**<br>Unavoidable and Unpleasant, but We Don't Have to Be Unprepared | 242 |

CONTENTS

| | | |
|---|---|---|
| 16 | **EMPOWERING**<br>The Gift of Opportunity | 256 |
| 17 | **COMMITMENT**<br>Be in It for Life | 272 |

NOTES 285
ACKNOWLEDGMENTS 303
INDEX 307

# LABWORK TO LEADERSHIP

# INTRODUCTION

"I WAS NEVER TRAINED FOR THIS." This thought was racing through my mind as I sat in my office grappling with how to manage the first major conflict in my research group. I was a few years into my faculty career, and a student in my lab had just completed their qualifying exam. They were angry about the feedback from their thesis committee and were letting everyone in the group know it. At the same time, the more senior students who had already passed their exams were frustrated by how the situation was distracting them from their research. Another faculty member on our floor had heard the rumblings of conflict and come to my office to let me know that if I didn't do something, my lab was at risk for a social implosion.

I remember my stomach filling with dread. I closed the door to my office, fought off the desperate urge to climb under my desk and hide, and proceeded to google the term "conflict resolution" in hope of finding some guidance to carry me through the situation. Not exactly the ideal on-the-job training scenario. Sound familiar? Maybe your issue wasn't conflict resolution, but if you are a faculty member, the chances are good that you've found yourself in more than a few situations where reality strikes and you suddenly realize that you were never trained for this job either.

How can this be? If you're like me, you walked into your office on the first day thinking, "Everything in my education has led up to this. All of my hard work as an undergrad, grad student, postdoc—*all of that has prepared me for this job.*" While this notion may seem laughable now, it was very reasonable then. After all, it's what most of us were told throughout our training, that our experiences were equipping us with the skills we would need for our future career. So, what are we missing? To answer that question, we have to consider how we got here.

As a graduate student, your days were likely focused on doing research, more research, and even more research, perhaps along with taking some classes, drafting a few manuscripts, or serving as a teaching assistant. This work carried on into your postdoctoral life, along with mounting pressure to secure a job and the realization that you also needed to generate your own research ideas and develop those to the point that a university or research institute would be convinced to hire you. Along the way, you may have also attended a workshop or two on pedagogy or mentoring to "prepare" you for your future as a faculty member. Then, it happened—you got the job! You may have walked in on the first day thinking that running your lab would be a research job, only to find your days filled with managing people, planning projects, resolving conflict, and navigating politics, all while trying to keep up with your email inbox.

This brings into focus a widespread challenge in academic science: *in addition to having a research job, you also have a leadership job, and it's likely that you were never trained for that.* Moreover, you were probably never even told that you would need such skills to be successful. It's not your fault that you're in this situation, but it is up to you to own the solution if you want to create a healthy work environment in which you and those around you can thrive. The good news is that you can do this, and this book is designed to help.

I know that you have walked a unique path to reach where you are now, and you have confronted challenges along the way that I likely haven't experienced. But if you've ever wanted to quit your job over the frustration of continuous rejections, or started your day with a ball of anxiety in your stomach (or tears streaming down your face) because of a difficult situation, know that you are not alone—I've been there too. I started my academic journey as a grad student and then a postdoc struggling with self-doubt so intense that it almost kept me from applying for faculty jobs. After a fraught search process, I managed to land my dream job and build a productive research program, only to find myself five years later facing an unfavorable tenure vote and thinking my career was over. I then moved to another institution and found myself confronted with intense bullying and harassment. Despite those struggles (and in some cases because of them), I am now a full professor with an endowed chair and I lead a thriving chemical biology research group at a university where I feel supported and valued. In addition to running my lab, I've had the opportunity to hold a variety of leadership positions within my institutions and the scientific community, and I'm currently serving as the chair of my department.

As I've navigated these struggles and opportunities, I have found tremendous power in applying principles and ideas from the leadership and psychology literature to the situations we face in academia. This began with the crash course in conflict resolution that I described at the start of this Introduction. In that moment, my focus was on simply getting enough information to formulate a plan for how to handle the conflict in front of me that afternoon. However, as I had time to reflect, it hit me that there is an abundance of resources out there for exactly the leadership challenges that I was struggling with—not only conflict resolution but also topics including motivation, dealing with failure and rejection, giving and receiving feedback, and fostering an ambitious yet collegial group culture. I started slowly, but before long I found myself devouring books and podcasts. Experienced leaders would probably view the things that I was learning as somewhat basic, but for me this new knowledge was transformative. More than a few times, I found myself listening to a podcast on the way to work and hearing exactly the information or advice that I needed to deal constructively with a situation that I knew I was going to face that day.

This raises a question: If it's that easy to find helpful information, why aren't we all basking in a wealth of leadership knowledge and insights? Perhaps the biggest challenge is time. Many of us feel like we don't even have enough time to do the things that are already on our list, let alone to read books or listen to podcasts across all of these topics. Another challenge is that most books on leadership are written for people who are running for-profit companies, not for academics pursuing curiosity-based research and focused on teaching and mentoring students. I would posit that we actually have more in common with CEOs than we might think, especially those of us who need to secure our own external funding and manage finances to ensure we can buy the lab supplies we need and still cover everyone's paycheck.

However, there are certainly many unique aspects to academia that can make it challenging to apply the principles and ideas from business management to our leadership roles in a research group.

My goal with this book is to distill the most important and relevant information from these resources and reframe it to be directly applicable to life in academia, so you can get the best possible return on your investment of time. I'll approach each leadership topic from the perspective of a faculty member leading a science lab at a research-intensive institution, while recognizing that your context may be different from mine. Perhaps you are a grad student, postdoc, or staff scientist who may or may not be aiming for a faculty career, but you have responsibilities leading and mentoring other researchers right now. Or maybe you lead a research group at a company or government lab. You may even be in the process of starting your own company. If you are a faculty member, your institution type and the norms of your research discipline may also create a unique set of needs. Although the examples I use throughout this book come from my own experiences, the underlying concepts can be applied broadly to help you grow as a leader wherever you are right now, as well as in the future.

I also know that, as a busy academic, you want a book that is a quick read and offers actionable advice, while as a researcher, you want to understand the evidence behind every claim. Thus, for each topic, I will spend some time diving into the literature and highlighting what researchers have found and what experts advise. Then I'll apply those principles to examples of real-life situations, with the goal of offering you something practical that you can use in whatever situation is right in front of you.

You might be wondering: How well did I manage that first conflict in my research group? Not very well. But I encourage you to also ask this question: How well did I manage the next conflict in my group?

Much better. What made the difference? Learning conflict-resolution skills. Even if you've never thought of yourself as a leader, you can learn to develop the skills you need to be an excellent one.

No matter what your research discipline or career context, there is one important factor that sets us up for success. As researchers, we are trained to be lifelong learners. Our job is to approach an area where there is a gap in knowledge, investigate and learn everything we can, and then communicate what we know to others. You are already primed to learn new things. You already have the mindset of doing this every day. I'm just encouraging you to aim that mindset in the direction of your leadership skills.

When we read the word *leadership*, we tend to think of it in the context of leading a group of people. This book starts out with an even more challenging person to lead—yourself. You may question whether developing your self-leadership skills will have an impact on the people around you. You can think of this like building a house, where the visible structure of leading others depends on the largely unseen foundation of leading yourself. Practicing better self-leadership can have a direct impact on those around you and even solve some of your other leadership challenges. For example, imagine what would happen if you improved your ability to cope productively with failures, such as a manuscript or proposal rejection. Chances are, this would also boost the morale and productivity of your research group, as you would be able to cast a candid yet hopeful vision for the future and then lead your team in that direction.

Once you're on the path to leading yourself well—having built habits for time management, goal setting, managing your motivation, and more—you'll be poised to think about your ideal workplace, and you may find that you have more power than you thought to create that ideal within your research group. We'll talk about how to make it a reality by taking your newfound knowledge and building on it with the skills needed to lead other people. It's been said that leadership would be easy if it weren't for all of the people. That is 100 percent correct. It's also correct that this job would not be nearly as fulfilling without all of those people. So, you need the skills to make the most out of your interactions with others.

We'll finish by exploring how to cultivate future research leaders. If our job was to maximize profit while retaining the best employees, we could probably do that by just being great managers. Our job is much more than that, however, as we are in the business of mentoring students and postdocs and leveraging the training opportunities in our research group to prepare them for their future careers. Some will go on to be faculty members themselves, and many more will end up as leaders in industry, government, or the nonprofit sector. What would it look like if we helped every individual to develop their leadership skills *before* they find themselves in a leadership position? We can do just that.

There is a final type of leadership that we won't address directly but that you'll find woven through many of the topics and situations in the following chapters, that of "leading up." This encompasses your interactions with anyone who is in a leadership role relative to you, such as your department chair, college dean, research center director, or funding agency program officer. While the specifics of managing those relationships effectively is a topic for another book, many of the skills we discuss will be directly applicable to your relationships with leaders above you and will help you have more positive interactions with them, as well.

As we take this journey together, I encourage you to remain authentic to your own personality and preferences. One thing I know is, if you're a faculty member, then nobody is going to tell you what to do or how to do it. I think that's part of what makes this the best job in the world, and I wouldn't dare to take that away from you. Thus, rather than telling you how to be a great leader, each chapter aims to get you started thinking about the topic and then provides you with information and examples so you can craft your own way of doing things.

When do we start? Today.

That reminds me of a conversation I used to have with my younger son. It went like this:

"Mom, when do we get to do that?"

"Tomorrow."

"Is it tomorrow right now?"

"Um, no. Unfortunately, the answer to that question will always be no."

While this memory still makes me laugh, I also recognize just how true this is for me when it comes to building new habits or changing how I do things. It's so easy to read something in a book and think, "That's great! I'll start doing that tomorrow." What I so often fail to recognize is that tomorrow never becomes today, and the days and months go by and I still haven't done the thing that I wanted to do. So, as you read through this book, commit to starting right now. Think about how you can put a new habit into practice today, even if it is just in a small way. Then keep doing that every day.

I'll even help you out with this. Just as I'm a firm believer that every successful meeting ends with action items, I end every chapter with a few suggested action items. You don't have to do all of them, or even any of them. But they will get you thinking about what you can do right now to take that step forward toward thriving in the leadership job that you were never trained for.

# I

# *SELF-LEADERSHIP*

# 1

## TIME MANAGEMENT

## Doing What's Important and Keeping It Organized

"HOW MUCH OF YOUR ABILITY to accomplish your goals will come down to the way you spend your time every day?"

That was the reply I received after I had emailed my friend Troy Champ to ask for advice about goal setting. Stunned by this question, I sat there silently for at least two minutes. As I stared at my computer screen, what sunk in was the simplicity of the question, the obviousness of the answer, and the realization of just how far I was from following the self-leadership principle that was implicit in all of it. While I had been focusing significant effort on setting goals, I was spending relatively little effort managing my time so that I could actually achieve those goals. I suddenly realized that, similar to online shopping, I could load up my cart with as many goals as I wanted, but paying for it all at checkout would come down to how effectively I was budgeting my time on a daily basis.

If you're reading this book, you almost certainly have some big goals. And you may also find your days consistently disappearing into a vortex of emails and to-do lists. Do you ever get to the end of a day or week and realize that even though it felt chaotically busy, you struggle to name one tangible thing that you accomplished? I'm going to ask you the same question that my friend asked me: *How much of your ability to accomplish your goals will come down to the way you spend your time every day?*

Let's take a look at why time management is so challenging. Your job comes with a large number of things that you either need to do or can choose to be involved in. This is arguably one of the great qualities of the job. The downside is that this means a large number of things are also competing for your time. Between handling the day-to-day logistics of a lab, mentoring the people in your research group, teaching classes, serving on committees, and all of the other specific responsibilities you have, it's a lot. I'm guessing that you've never been given the opportunity to take a class on time management, or if you have, that you still have more to learn. The good news is that, much like with managing your money, there are some simple strategies you can implement to more effectively manage your time and ensure that you are spending it in a way that helps you achieve your goals.

## Be Kind to Future You

Let's start by going to a key source of the problem. There are certainly many things that we absolutely have to do in order to be successful. But all too often we agree to do things that we don't actually have to do—or want to do. This could include the committee you agreed to serve on, the paper you agreed to referee, the new collaboration that sounds great but you know will never be funded or is just not a top priority. If you aren't far enough into your academic career to have experienced these specific situations yet, you have probably agreed to do something in another area of your life that you didn't actually want (or need) to do.

Why do we do this to ourselves? There are many answers to that question, and none of them are particularly encouraging. We may say yes out of guilt or fear, feeling that we owe it to someone. If you are early in your career, this can be particularly challenging when the person asking has more power than you and may even vote on your tenure or promotion in the future. (I'll say more later on how to navigate this power dynamic.) Sometimes decisions can be driven by arrogance, when we think that the task won't get done "right" if we're not involved. Finally, there's emotional laziness—that feeling that it's easier to just say yes than to do the emotional work of disappointing someone. I fall into all three of these traps on a regular basis. Although the reasons we end up agreeing to do something when we really should decline can vary, one common thread is the relationship between "current you" and "future you."

According to a principle called future self-continuity, we tend to feel relatively little connection to the future version of ourselves.[1] This means that, although you are probably a very delightful person, current you can be less than kind to future you by agreeing to take on too many things. In more scientific terms, future self-continuity has been linked to the phenomenon of temporal discounting, in which we give less weight to the consequences of decisions that will have an impact on us in the future compared with decisions that will affect us immediately.[2] Temporal discounting has been observed in a wide range of decisions, from choosing what percentage of our salary to set aside for retirement to the likelihood that we'll exercise on a regular basis.[3]

In his book *The Productivity Project*, Chris Bailey describes how temporal discounting can also have an impact on time management. When you're asked to commit to doing something in the future, current you has very little empathy for future you. Unpleasant commitments are added to your calendar when current you agrees to do things that future you won't actually want to do. And the further into the future those obligations are, the greater the effect of temporal discounting and the more likely you are to say yes to them. In fact, I've found that even when the date of that event or deadline finally rolls around and current me is cursing the past me who put that task or meeting on my calendar, I still have difficulty learning from the experience and changing my habits.

How do we break this cycle? In one study involving exercise, researchers found that when participants wrote a letter to their future self, it improved their future self-continuity and led to an increased frequency of choosing to exercise.[4] It's probably not practical to sit down to write a letter to yourself every time you need to make a decision about whether to commit to a relatively small task, such as a manuscript review. But we can still harness the principle to make better decisions. When you are asked to commit to something that won't require your time or effort until a date far in the future, take a moment to visualize that future version of you. What will you be thinking or feeling when you walk into work and have that task on your to-do list? I've also found it effective to ask myself, "If I had to do this tomorrow, would I say yes?" I imagine myself having to immediately add it to my to-do list or shift my schedule around to prioritize the task. This can help overcome temporal discounting and put my focus on what I would actually be committing to and whether it would be a wise use of my time. I also use positive reinforcement to reward myself for making good time-management decisions. When I say no to something that I didn't really want to do, sometimes I still put it on my calendar but with a note saying that I've declined. When that date rolls around and future me sees what past me has saved myself from having to spend time on, I give myself a high-five through time.

## Every Yes Is a No

The essence of the time-management challenge is that every yes to something is a no to something else. Time is finite, and so whenever you choose to spend time on an activity, there are a variety of things that you're no longer going to have time to do. Sometimes that is a wise choice—for example, deciding to work on a grant proposal, even though it means not having time to sit on the couch and binge-watch Netflix. Or making time for a morning walk or yoga session even if it costs you the opportunity to press snooze or stay in bed for an extra half hour to scroll through social media. For many years, my approach to making decisions about whether to do something boiled down to, "Well, I don't have anything on my calendar then, so sure—I'll do it." I still fall into that trap at times, but I've gotten better at thinking through another question: "If I say yes to this, what will I have to say no to, and am I willing to make that trade?" The recognition that every choice about how I spend my time is a trade-off has proven to be transformative for my career.

This all sounds nice in theory, but what do you do when you're in the moment and the guilt or sense of importance is weighing on you? At those times, you need to make the other side of the equation as compelling as possible. What is the thing that you most love doing but haven't had time for lately? When an opportunity comes along that seems important but you know you probably shouldn't commit to it, ask yourself, "Would I rather do this or have that time for [the thing I really enjoy but never get to do]?"

You can also slow down the narrative and give yourself time to carefully consider what the commitment will cost you. In *The Coaching Habit*, coaching expert Michael Bungay Stanier suggests that sometimes we just need to "say yes more slowly." Rather than committing to something immediately, Stainer suggests pausing and asking questions to figure out what would be involved, how much time it would take, and whether the person asking would be willing to take something else off of your plate to free up time for the new task they want you to commit to. In the case of an academic committee assignment, for example, this could mean asking how often the committee will meet, how much work would be required of you outside of meetings, and whether you could be released from some other service activity to make the time for this one.

## Choosing Your Yes

Up to this point, I've been almost exclusively talking about how not to agree to new opportunities. That's because in general we err on the side of overcommitting. How many faculty members do you know who undercommit? Maybe a few, but let's be honest—they're probably not reading a book about leadership right now. While it's important to say no to some things, there are many things that are not required but we absolutely should say yes to. How do you figure out which things belong in which category?

This brings us back to the story at the start of this chapter. Asking me, "How much of your ability to accomplish your goals will come down to the way you spend your time every day?" was a brilliant move on the part of my friend. Instead of telling me what to do, he had posed a pointed question, knowing that once I answered, I would be compelled to act. Without writing it out, what he was really asking me was, "Now that you've admitted that your time-management skills are holding you back, what are you going to do about it?"

So, what did I do? I designed a priorities exercise that helped me create a decision-making rubric. Here's how it works.

*Step 1.* Make a bullet-point list of all of the different tasks that you spend your time on. You'll find that this list quickly becomes very long as you factor in things like teaching classes, editing manuscripts, reviewing grant applications, meeting with lab members, attending committee meetings, booking travel, meeting with seminar speakers, traveling to conferences, and so on. Be sure to also include the things you spend time on that are not work related—exercise, hobbies, time with family. When you think that you've got everything on the list, take a moment to look back over your calendar (and your to-do lists, if you have them) from the past two months. If you're like me, this will remind you of another ten things you need to add to your list of tasks.

*Step 2.* Define two to four groups of stakeholders. These are the people for whom you do all the things you do. As a faculty member leading a lab, I broke down my stakeholders into three groups: my research group, my academic community, and myself. This also highlights one of the rather unique things about this job—running a research lab is essentially a full-time job, but that group of stakeholders constitutes only a fraction of the people for whom we need to do things. In my academic community I also have stakeholders in my department, the university, and the broader research community. And don't forget about commitments outside of work. I decided to put friends and family in the "myself" group, but you can create a separate group for them if you prefer. Once you've defined your stakeholder groups, color code or make a note next to each task to indicate which group it is for.

*Step 3.* Work through your list of tasks and categorize each one as essential, exceptional, or neither. Depending on your job, there may be relatively few tasks that are truly essential, so you can define this category as those things you are expected to do if you want to be considered a high performer in your research field or department. You can also place in this category activities that you consider nonnegotiable for maintaining a work-life balance and your own well-being. Exceptional activities are things you do that you aren't required to do, but that you are really good at or you especially enjoy doing. As we'll see in Chapter 2, these are the tasks that leverage your unique strengths—that make you feel like you really stand out and contribute in a way that others might not be able to. These tend to be the tasks that you enjoy the most, as they help you feel creative and, well, exceptional. Be discriminating in doling out the "essential" and "exceptional" tags—you should finish this exercise with several things left in the "neither" category.

*Step 4.* Create a grid in which the columns are populated with your stakeholder groups and the rows are titled "Essential," "Exceptional," and "Say no to more often." Cut and paste each of the tasks from your list into the appropriate box, based on the category assignments you made in Steps 2 and 3.

This is now your rubric for figuring out when to say yes and when to say no. If a task is in the essential category, yes, you need to find time for it. Hopefully you have included in this group both work tasks and self-care activities like exercise and hobbies. Later in this chapter I'll talk about why self-care is important and how to make sure you create time for it. If a new task you're considering is in the exceptional category, it is something you should do whenever possible. This is when it can be helpful to think about the yes/no trade-off. Ask yourself, "If I commit to this activity from my exceptional column, is there another task in this column I would be willing to drop?" This brings us to the final column—things that you should say no to more often. If a task is in this category, then you should do exactly that. These tasks aren't necessarily bad things to spend time on, they're just bad things for *you* to spend time on, because they don't leverage your unique skills. And you don't have to say no to them all of the time, just more often than you currently do.

Referring to this grid helps me prioritize my daily tasks, and even more important, it has saved me from making significant commitments that I would have regretted. For example, being a journal editor is one task that fell into the "say no to more often" category on my priority grid. For some, serving as a journal editor is an important step in their professional advancement. It is also a role that many researchers enjoy, and a way for them to make a significant contribution to their field. However, I discovered that for me it is neither essential nor exceptional, and that committing to that task would cause me to miss out on doing other things I view as more important, like writing this book. Despite knowing this, when I was recently asked to become an associate editor for a new journal in my field, I desperately wanted to accept the opportunity. It was interesting and important work, and let's face it, who doesn't love the prestige and affirmation that comes with such a position? Deep in the process of selling myself on saying yes, I was only able to rescue myself from that sales pitch by going back to my priority grid and reminding myself that this was not something I should be committing that much time to. Even with this reinforcement, it was still very challenging to decline, but I'm also very glad I did. Owing to all of the tasks that are now *not* on my weekly to-do list as a result of that decision, I have time open for other things that make me more effective and make better use of my unique talents and skills.

It's important to note that your priority grid doesn't have to be static, as activities can move between categories over time. Organizational experts Amy Wrzesniewski, Justin Berg, and Jane Dutton have popularized the concept of "job crafting," which involves interrogating the individual tasks within our current job and adjusting how we spend our time to emphasize tasks that align with our motives, strengths, and passions.[5] Even if you like your job, you may hit a point where you start to feel stuck or in a rut, and your priority grid can provide a starting point for thinking about what tasks you might be able to move between categories in order to break out of your routine, develop a new skill, or maximize your impact.

## Perfecting Your No

Okay, we've taken a huge step forward—you have a list of the things that you should start saying no to. Now, how do you do that? If you're a people pleaser like I am, an essential first step is letting yourself off the hook. Recognize that your reason for declining is valid and not selfish, even if it is simply, "I don't have enough time right now."

The next step is to consider the power dynamic between you and the person you need to say no to. If they have significantly more power than you do or you have relatively tenuous job security, or both, then you may want to enlist a mentor or advocate who can decline on your behalf. Perhaps you find yourself in your second year as an assistant professor, and after being given a one-year pass on joining thesis committees, your department's director of graduate studies (who is a tenured senior faculty member) has asked you to join the committees of ten of the first-year students. If this does not feel like an equitable or reasonable request but you don't feel comfortable pushing back yourself, then another senior colleague can be an especially good ally. You may not even have to ask directly—just a comment like "I was surprised to be asked to serve on so many more committees than the other faculty in our department" may spur your colleague into action on your behalf.

For the times when you are in a position to decline directly, there are a variety of formulas for how to say no, and you may already have something that works for you. My approach is fairly direct, as that is what I appreciate when I'm the one whose invitation is being declined. First, I express my appreciation for the opportunity. Then I state why I can't commit to the thing that I'm being asked to do. Some people would argue that you don't owe anyone an explanation, and that's true. Still, it is something that I value as a person and so I nearly always include one. If the thing I'm being asked to do directly conflicts with something already on my schedule, then I just say that. It gets more complicated when it doesn't technically conflict, but adding it to my schedule would take away the time I need to do something else. Basically, I know that if I committed to the new task, I would end up doing both things poorly. In these cases, I explain that I already have too many commitments and I wouldn't be able to give their task the level of attention they would want me to. I've yet to have someone reply, "Please say yes anyway and do a rushed and careless job." Finally, I offer an "alternative yes." This can be as simple as saying, "Please do keep me in mind for future opportunities." This three-part response formula has given me a way to simultaneously honor my values, my relationships, and my time. Take a moment to think about a formula that could work for you. Maintaining your mental health and work-life balance means saying no to a lot of good opportunities, and it will be much easier to do this if you have already planned your response.

## Learning to Juggle

What about all the things that are constantly coming out of nowhere and requiring your time? Sometimes these are fun interruptions, such as when a student in your group stops by your office to discuss some exciting new data. Other times they're less fun, such as when you receive an unexpected request for additional compliance paperwork from one of your funding sources. Either way, the impact on your schedule creates a real challenge. One thing you can do to help manage these tasks is to build margin into your schedule. This margin is time that you block off in advance even though you're not sure yet how you're going to use it. When unexpected requests or emergencies pop up, then you have that margin built in, allowing you to enjoy the fun interruptions and feel less stressed by the unpleasant ones.

Speaking of unexpected requests, one of the biggest complaints I hear from colleagues is how difficult it is to handle the influx of email they receive. This is a critical task to manage, and I think we could all agree that email is a universal source of stress. This is not surprising if we think about how it works—email is designed such that at any moment, someone can push their way into our schedule and demand our time or attention, sometimes immediately. Nevertheless, we can manage this so that it doesn't consume our day or drive our stress level through the roof. We can be intentional in deciding whether we want to keep a consistent eye on email or check it only at specific times of the day. If we set aside time for the inevitable work that email requires, we can reduce the stress associated with the constant deluge of messages in our inbox.

You might be thinking, "This all sounds great, but what if I have so much on my plate that I can't set aside the extra time? What happens when something comes out of nowhere and requires much more time than what I've planned for?" I understand. Despite my best attempts at time management, I still have days, weeks, or even months when I just can't keep up. In these cases, it's important to be actively reprioritizing tasks on a daily or weekly basis.

Former Coca-Cola CEO Brian Dyson introduced the metaphor of juggling with rubber balls and glass balls—sometimes you have to let a ball drop to the ground, and it's important to choose a rubber ball that can bounce back up instead of a glass ball that will shatter on impact. Dyson introduced this concept in the context of choosing family, friends, and health over work, and it can also apply to the decisions we make within our work. If you are overcommitted and know you are going to have to miss a deadline, recognize which tasks are rubber balls and which are glass balls. A manuscript review is almost always a rubber ball—if you miss your deadline, it will bounce back again and again as you receive reminder emails from the journal editor. A grant deadline for an agency that accepts proposals only once per year is a glass ball. There is no easy fix for that one if you let it drop.

## The Most Important Priority

Your health is another glass ball. When we're physically tired, it's difficult to get out of bed, get into work, and have the stamina to get through a full day. When we're mentally tired, it's difficult to get motivated to do our work and find the inspiration to do it well. When we're emotionally tired, we have no margin left for navigating challenging conversations and supporting those around us. Collectively, this means that when we aren't healthy, it's nearly impossible to function at our best and be an effective leader. All of the time-management strategies in the world won't do us much good if we aren't making time to take care of our physical, mental, and emotional health.

I'd guess that there are few of us who don't at least *want* to make time for self-care activities such as exercise, hobbies, or relaxation. We start each day with the best of intentions but then get bogged down by the flood of email messages or pulled into an unexpected meeting. As we struggle to reprioritize so we can get everything done before our deadlines, often our self-care activities are the first thing to be jettisoned. No matter how much we plan, we all occasionally have days when that happens. In the short term, the impact of missing that grant deadline may be greater than the impact of missing a woodworking class. But it's important to recognize the cumulative impact of not prioritizing self-care and to make sure that in the long term, this is the exception rather than the norm.

There are several things you can do to maintain consistency in your self-care activities. The first of these is to have a specific plan. If all you have is a vague notion that you want to "work out" or "go relax," then the uncertainty will turn it into a task that you easily procrastinate on or skip entirely. What exercise or self-care activities do you enjoy? Are there goals associated with them, like getting in shape to walk or run a 5K or finishing a book you've been meaning to read? Having a specific goal helps you to craft a plan for exactly how you want to spend your time. Then put that time in your schedule. I know that if I start my day thinking, "I'd love to go swimming today—I'll do that when I feel like I have some free time," the chances that I will make it to the pool are exactly zero. Instead, I look at my calendar, choose a time, and schedule it in. Then when that time rolls around, it is easier to set other tasks aside and stick to my plan.

I've also intentionally arranged my commitments to include other people, which makes it harder to back out of them. For several years, I ran at the track every Tuesday night at 7:00 p.m. I was part of a running group, and we had a coach who would create our workout goals for us. I rarely missed this run unless I was traveling. Why? First, the activity happened only once each week, and I knew that if I let other things get in the way, then I wouldn't get another chance to run with the group until the next week. Second, I had friends in the group I genuinely looked forward to seeing and who might call me out if I didn't show up. Third, I was paying money to participate, and so I was financially invested in the activity. If group activities don't align with your preferences or schedule, you can still think about how to create structures that encourage you to save time for the things you want to do. For example, if you live relatively close to work, you might consider intentionally not purchasing a parking pass so that you will have to make time to bike or walk to work each day. Whatever your context, the key is to think about ways that you can build safeguards into the activities that are important to you.

If you don't believe you have enough time in your calendar to schedule activities for self-care, it may be time to reevaluate your priority grid and try to add more tasks to the "say no" column. In her book *Drop the Ball: Achieving More by Doing Less*, leadership expert Tiffany Dufu advises that "doing it all" is an impossibility, and that embracing imperfection can actually be the key to "having it all" because it creates the conditions we need to thrive both personally and professionally. There may be some balls that we drop and never pick up again, Dufu says, and rather than feeling guilty for cutting those commitments from our to-do list, we can view these choices as intentional steps toward achieving our goals.[6]

Along these lines, while it may seem counterintuitive, creating even small spaces in your schedule for things like exercise can actually help to make your schedule more manageable. A 2008 study showed that workers who incorporated exercise into their daily routine experienced increased productivity, which the researchers primarily attributed to the workers' improved mood.[7] In a more recent study, sports-science researcher Maria Küüsmaa-Schildt and colleagues explored the relative effect of morning versus evening exercise. They found that both led to similar increases in physical strength and self-esteem and that neither had a negative impact on sleep.[8] In fact, carving out time for activities such as exercise can actually save you time in the end. Sure, there are limits to this. If your goal is to train earnestly for an Ironman Triathlon, that will probably require some major restructuring of your work and personal commitments. However, I would guess that almost all of us can find thirty to sixty minutes a few times per week. In the case of exercise, that is more than enough time to generate real benefits, including feeling better about yourself, increasing your energy and focus, and investing in your long-term health.[9] If you're skeptical, then do the experiment—try scheduling one hour for exercise or a hobby four times per week for four weeks, stick to your plan, and see how it goes. If you look at your calendar and you can't find that time right now, keep scrolling further out until you do find it. Put it on the calendar now and it will be waiting for you in a few months.

## Keeping It All Organized

You've decided what to say yes to and what to decline. You know which things need to be prioritized and which things can occasionally slide. Congratulations! That's a huge step forward in managing your time. However, you will still need a way to organize and keep track of all of these things on a daily, weekly, and monthly basis. The best system is the one that works best for you, and it's likely that you already maintain some form of a to-do list and a calendar. I would encourage you to connect them by regularly converting your to-do list into a schedule built around your calendar. As you make your list of tasks that need to get done, outline how long each one is going to take and make a plan for when you will work on it. Why does this matter? If you're like me, you can fall into the habit of spending tons of mental energy thinking about what you need to do and worrying about whether you will have enough time to get it all done. But if you're able to look ahead and map out not only what you need to do each day, week, or month but also when you will do each thing, that can relieve much of the mental burden.

Every person is different, and you will need to experiment to figure out what approach to planning works best for you. I personally like to map out my plan for each week at the end of the previous week. As I head into the weekend, I find that a key to ensuring that I can truly relax is taking a few minutes to look over my to-do list for the upcoming week and create a plan for getting it done. This helps me feel assured that I can put away my laptop for a day or two and my world will not be a raging dumpster fire of impossible deadlines when I return to it on Monday morning. So, I sit down with my list of tasks I need to complete in the next week, and I also pull up my calendar. I estimate how long each task will take and then schedule it into one of the windows of free time in my day. I make sure I also schedule the times that I need for self-care activities and that hour each day for managing email. Some people like to write their tasks on their calendar as if they were meetings or appointments. I personally prefer not to build in quite that much structure, as I value the freedom to spontaneously choose in what order I'll tackle my to-do list each day. What's important is that I've decided what needs to get done in the upcoming week, I know that I should have time to accomplish those things, and I have a plan for how that will happen. This allows me to let go of the anxiety and enjoy some rest and self-care so I'll be at my best when Monday morning rolls around.

## ACTION ITEMS

- *Set your priorities.* I hope you took a moment to complete this exercise as you read the chapter. If you did, congratulations—you've already completed the first action item! If you didn't—ahem—have time, then take a few minutes now to complete it.
- *Create your "no" template.* Look back through the emails you've written in the past to decline participating in something. Choose ones in which you really wanted to say yes but just didn't have the time for the task. Use those messages along with the formula outlined in this chapter (or another formula that works for you) to create a template for a "no" email and save that file on your desktop or somewhere easy to access. Use it frequently.
- *Schedule your self-care.* Choose one self-care activity that you either are doing or want to start doing and put it on your calendar on a recurring basis. Even better—do the experiment and set aside one hour, four times per week for four weeks, and see what happens.

# 2

# LEADERSHIP STRENGTHS AND STYLES

## What Makes You Weird?

BUSINESS CONSULTANT MARCUS BUCKINGHAM SAYS that we need to understand what makes us weird. If our default mode is to approach leadership assuming that everyone else is just like us, then we will end up doing things exactly like everyone else does. At that point, we're just one small step away from the dreaded "that's the way we've always done it." If we want to break out of this homogeneous way of thinking and leading, we need to understand what makes us unique. During an interview on Adam Grant's *WorkLife* podcast, Buckingham says, "I think, deep down, we don't actually think that our unique way of engaging with the world is worth uncovering. And what we need sometimes is someone going, 'No, no, no—that's you. That's weird.'"[1]

Compared with most careers, being a faculty member can offer tremendous freedom to express our uniqueness. There are certainly things that we have to do, and some of those things need to be done in a very specific way (think: conflict-of-interest disclosures or quarterly effort reporting for federal grants), but for the most part we have wide latitude. Think about all of the different ways that you could design a course, structure a research project, or organize the work of a committee. The possibilities are endless. One of the places where this freedom is most important, yet often underutilized, is in managing our research groups.

Every year, my lab members and I organize a group retreat. Sometimes we rent a cabin in the woods and sometimes we just find a conference room in our department. No matter our location, we set aside our experiments for two to three days and devote this time to reflecting on our accomplishments from the past year, setting goals for the coming year, brainstorming on projects that we are planning to pursue, and strategizing new ways to organize our lab. The exact agenda changes every year, depending on our lab's needs, but it always includes some breaks from the work so that we can get outside and enjoy nature or laugh together over silly card games. Over dinner at a conference a few years ago, I found myself describing our retreats to another faculty member. They asked me where I found the instructions for how to run a retreat and who told me that I was allowed to do so. My response was, "We don't have instructions. We just think about what we want to accomplish each year and then create a schedule around that. And nobody ever told me I could do this. I do it because nobody told me that I couldn't." The look on my colleague's face said "that's weird," and I took it as a compliment.

For you, organizing a retreat for your lab may sound like fun, or maybe it sounds like a total nightmare. Either way, the point here isn't that you should be holding a retreat for your group every year, but that you should be leveraging your natural strengths to think about creative ways to lead your lab. You may have the impression that leaders need to have specific characteristics, such as being extroverted and charismatic, and there is good reason for that—a 2009 study by psychology researchers Deniz Ones and Stephan Dilchert showed that 96 percent of managers or executives in industry leaned toward having an extroverted personality. At the same time, some leadership experts are pushing back on this bias and advocating for a "quiet revolution," in which the strengths that introverts bring to their leadership are also recognized.[2] If you take a moment to think about the leaders that you look up to and respect, perhaps a department chair, former research advisor, or colleague who serves as a mentor to you, you might realize that you already know several introverts who are outstanding leaders.

Perhaps you lean toward being introverted yourself, and large group activities push you out of your comfort zone. Leveraging your strengths might instead look like partnering with each of your lab members to formulate and check in on their individual development plans. Or perhaps your strength is in the group culture that you foster by creating fun traditions to honor your group members' publications and successful milestone exams. When it comes to our research, we know that we need to be innovating constantly and that repeating the same things we did five or ten years ago will not lead to success. So, why don't we apply the same mentality to our leadership? Why don't we spend more time experimenting and innovating in how we manage our labs? To innovate most effectively, you need to uncover what makes *you* "weird," and that means taking time to intentionally assess your strengths.

## Ignoring Your Weaknesses

As you finished reading that last sentence, there's a good chance that your brain interjected with the words ". . . and weaknesses." If you're like me, you were taught that we have both strengths and weaknesses, and our goal is to maximize our strengths and fix our weaknesses. Leadership development consultants John Zenger and Joseph Folkman disagree. In their decades of research encompassing tens of thousands of leaders, they have found that unless a leader is struggling with a "fatal flaw," they are more likely to succeed by capitalizing on their strengths than by trying to address their weaknesses.[3] Zenger and Folkman define a fatal flaw as something so significant that it is likely to derail the person's career or impede their success despite anything else that individual might do. They're talking about things like an uncontrolled temper that leads to frequent angry outbursts or a lack of integrity that causes a leader to completely lose the trust of their team members.[4]

Marcus Buckingham takes a slightly different approach. He argues that strengths are not good or bad, but are simply natural traits that we can either use well or use poorly. Much of the advice we receive about trying to fix our weaknesses is essentially telling us to be less of who we naturally are, he explains, and that's kind of impossible. Rather, we should focus on identifying our natural strengths and preferences, and then figure out how we can use them most effectively.

What does this look like in real life? I'll give you one of my examples. I am someone who loves to get things done. Crossing things off of my to-do list brings me great joy, and I want to experience that dopamine rush as often as I can. This personality trait often helps me as a faculty member, as I can move through different tasks quickly and I'm willing to accept when things are "good enough," rather than getting hung up trying to achieve a level of perfection that doesn't exist. My rush to get things done can also work against me as a leader. If I'm meeting with my research group to strategize about how we're going to run a set of experiments, I will feel tempted to make a decision as quickly as possible so that I can check it off and move to the next thing on my list. But if we rush through and make a poor decision, it could end up costing us weeks of progress on the project and hours in future troubleshooting meetings—a net loss in the economics of time and finances. It's always going to be my natural inclination to want to do things as efficiently as possible, but I've learned that there is a time to push ahead quickly and a time to slow down to make sure we get it right the first time.

## Gathering More Data

Perhaps you already have an idea of what some of your unique strengths are, but maybe you are struggling to figure all of them out. There are times when you or someone around you realizes that something you do naturally is unique to you. You can also take notice of the times that you feel most fulfilled and competent—when you're "in the zone"—and then identify what you're doing and what aspects of your personality you are getting the chance to use in those moments. Beyond these serendipitous discoveries, it can also be useful to seek out this information in an intentional way by using a personality assessment tool.

Depending on the experiences you've had with these types of exercises in the past, you might be groaning at the thought of a "personality assessment." Stay with me. A key purpose of the assessment is to give you a framework and a language for thinking and talking about who you are and what it looks like when you use your strengths well. It can help you to see actionable ways to be more effective in leading yourself and leading others and how you can use your unique characteristics to do something that other people might not be able to. It can also help you to see the ways in which your strengths can work against you if they are not used well, and how to guard against that happening. Finally, assessment tools give you a common vocabulary for discussing this with the people around you, and in Chapter 10 we'll see how that can be especially useful in conflict resolution (or even better, in conflict prevention). If you're still skeptical, I encourage you to view this as research. Here is another opportunity to collect and analyze data about an exciting topic—you!

There are several assessments to choose from, and so there's a good chance that you can find one that is helpful and makes sense to you. You might question whether any test can really capture the vast complexity of what makes us who we are, and that is a fair point. Moreover, your personality traits can change over your lifetime as a result of your experiences.[5] But the utility of these assessments doesn't rely on their being perfect. A famous saying in statistics is that "All models are wrong, but some are useful." This is true for personality assessments. They won't be able to tell you everything about who you are as a person, but they can help to illuminate some of your unique strengths and characteristics. You could argue that you already have a good sense of your strengths—after all, you've lived with yourself for many years now. This is also true. Many of the results of a personality assessment will not surprise you. Still, some might.

There are far more assessment tools out there than I could list (or than you would want to read through), so I've chosen and highlighted a few of the more popular frameworks and tried to cover a range of costs and levels of complexity. I tested each of these out myself, and here's what I found:

## Big Five Personality Test

The Big Five is one of the most well studied and validated personality frameworks, as it has been applied to everything from academic performance to financial decision-making to social media use.[6] As the name implies, it is based on five fundamental personality factors: conscientiousness, agreeableness, neuroticism, openness to experience, and extraversion. For each factor there are two opposite poles and everyone lies somewhere on the spectrum between them.

*The survey.* Because the Big Five is such a widely studied framework, numerous personality assessments have been created that use this model. Many of them will give you a series of statements and then ask you to rank how well each statement reflects your personality. Several of these tests offer a free version, though you will probably have to pay something to get a version of the results that really digs into what your scores mean.

*The results.* The primary output you'll receive is a score indicating where you rank on each of the five factors. Since we're academics who think of percentile scores in terms of good or bad grades, it's important to note that a higher score isn't necessarily better. Rather, it just means that you more closely align with the personality trait that has been arbitrarily assigned to the high end of the spectrum. If you use a paid version of the inventory, in most cases you will also receive a report that describes how your unique combination of scores across the five factors shapes your strengths and preferences.

*How this is (and isn't) useful.* One of the especially useful aspects of the Big Five framework is that it has been extensively validated. Perhaps most encouraging is that multiple groups of researchers seeking to distill the primary factors that make up human personality have ended up with the same set of five traits. This means that the results you receive will probably reflect strengths that show up in your everyday behaviors and interactions with others. This tool can be especially helpful for teams of individuals, as knowing the ways in which you are similar to and different from the people you work with can illuminate potential points of conflict and help you to appreciate the contributions of different viewpoints and ways of approaching tasks. Where many versions of this test are limited is in the scoring, as you only receive a score across each of the five traits. Compared with some of the other assessments that integrate these traits into more nuanced categories, it can be difficult to see how that information might be useful in guiding your career choices or how you interact with others.

## Enneagram

The Enneagram framework dates back to the fourth century but was refined and popularized in the mid-twentieth century. Although researchers are only beginning to study its validity, Anna Sutton and coworkers have found that it has potential for use in development and management.[7] Similar to the Big Five, the Enneagram uses your self-reported preferences and personality traits to characterize you according to nine different interconnected personality types, such as "The Individualist," "The Investigator," and "The Challenger," represented by the points on the nine-sided figure known as an enneagram.

*The survey.* Several versions of the survey exist. Given the age of this assessment tool, no official organization owns it, but a number of companies offer paid versions of the survey that include a detailed report on what your results mean and how to use them. As with the Big Five, there are also several free versions available online.

*The results.* Your survey results will indicate which personality type of the nine options you most strongly align with. For each personality type, you can find a description as well as examples of what that set of characteristics looks like when someone is at their best or at their worst. As an example, my personality most strongly aligns with a type three, "The Achiever." This is reflected in my obsession with productivity—at my best, I set ambitious goals and encourage the people around me to do outstanding work, but at my worst I can be insecure and hypercompetitive, envying the success of others.

*How this is (and isn't) useful.* While the Enneagram has fewer "types" than all of the possible combinations in the Big Five, I've found it to be more useful. The personality traits that it measures are more variable among academics, and thus it more effectively highlights different perspectives and ways of thinking among the people I work with. One downside of the Enneagram is that it has not been rigorously validated, and this is further complicated by the many versions of the survey that are available. Nevertheless, there is strength in the simplicity it offers—even though it isn't perfect, having a simple language of nine types as a framework for discussing differences in work or communication styles can be highly useful. It also offers a convenient way to think about which people best complement our strengths. For instance, my personality type values productivity, even when that means occasionally cutting corners. This is useful for tasks that just need to be "good enough." When the task is critical and I need to slow down, I can make sure there is someone in the room who is an Enneagram type one ("The Perfectionist"), as that person will pay attention to detail and insist that our finished product be as good as possible.

## CliftonStrengths

More popularly known as the StrengthsFinder test, this assessment tool was developed based on research by Marcus Buckingham and

Donald Clifton.[8] In line with Buckingham's philosophy, which I described earlier, this assessment was among the first to focus specifically on strengths and to advise people to not spend much time or energy worrying about their weaknesses. The framework looks at thirty-four different strengths that can be divided into four categories: executing, influencing, relationship building, and strategic thinking.

*The survey.* The survey consists of 177 questions, and you have only twenty seconds to answer each one—the aim being to garner your gut responses (and also enable you to finish answering all of the questions in a reasonable amount of time). Each question gives you two traits, and you rank on a five-point scale where you identify between those two traits. While there is no free version of this assessment, the basic test is relatively inexpensive. There are more expensive options that include an analysis with a certified coach.

*The results.* The report you receive highlights your top five strengths and how you can make the most of them in your work and your interactions with others. The strengths ranked six through ten also receive some attention, though less than the top five. Strengths not in your top ten are not viewed as weaknesses. Rather, they are things that you just don't naturally do. The goal isn't to remediate them but to accommodate them. For example, if adaptability ranks low among your strengths, you may struggle in situations where the plan changes suddenly or your schedule is frequently interrupted by emergencies you need to deal with. If you try to completely change who you are to be a "go with the flow" type of person, the result will almost surely be frustration. Instead, you can look for roles that don't require this strength, for example by chairing an undergraduate curriculum committee working on a long-term project rather than being the safety officer, which involves dropping everything whenever there is an accident or a hazardous situation in your department. You can also be mindful in the unavoidable instances when this strength is needed, such as when a member of your lab walks into your office to share that they just lost a close family member. In such situations you can experiment with different approaches, such as pausing to take a deep breath or to shut down your computer, and using this moment to accept the fact that your plans for the next hour have just changed. Even acknowledging to yourself that unexpected situations are not your strength can help you adapt to the situation and do your best, in this case, as you try to support your grieving lab member.

*How this is (and isn't) useful.* Compared with the previous two examples, this framework offers a much finer level of detail. Personality traits are broken down into thirty-four different strengths, and the assessment report creates a unique picture of your own combination of top strengths. I also appreciate the underlying message that there is no single set of strengths that makes a "great leader." Instead, the framework emphasizes that we each have the ability to lead well in our own unique way. The main downside to this test is that there are no free versions, so you have to spend money on something that you may or may not find useful.

## The Birkman Method

This is one of the most complex frameworks, as it offers users insight into their overall personality type, usual behavior, needs, and stress behaviors along a variety of metrics, as well as their interests and decision-making styles. At its core, the Birkman Method assessment is based on the idea that the things we think other people want or prefer actually reveal our own needs and preferences.[9]

*The survey.* The Birkman Method's survey consists of a series of true-or-false questions followed by a series of rank-order questions.

This is by far the most interesting of the surveys, and I think that is part of its power—as you complete the survey, it's not uncommon to find yourself wondering, "Why in the world are they asking me this?" With other surveys, you can generally tell in advance how your answers are going to influence the results, but this survey is almost impossible to game. Also, unlike the other assessments, the Birkman survey must be administered by a certified professional. Thus, unless your institution already has someone who is certified in the Birkman Method, this may be the most expensive of the options.

*The results.* As I mentioned, the survey results are multifaceted. The Birkman Map is a two-dimensional grid that uses your responses to locate you along axes ranging from introverted to extroverted and task-focused to people-focused. Your normal behavior and stress behavior are mapped onto the resulting quadrants. Your results also give you a number from 1 to 99 for a variety of metrics, such as assertiveness and physical energy. As with the other frameworks, no score is better than any other. They just tell you something about how you are wired. What is unique about the Birkman Method is that for each characteristic, you receive three scores: "usual," "needs," and "stress." These are pretty much what they sound like, but very powerful. When you are in an environment that accommodates your needs characteristics, your usual characteristics describe how you will generally behave. When your needs are not met, your behavior changes and reflects your stress characteristics. As an example, for the metric of physical energy, I am a usual 97, with needs and stress both at 2. This means that I usually bring a lot of energy to my tasks. But when other people step in and demand even more from me, this goes against my needs, and I switch to my stress behavior of shutting down and retreating. The Birkman report also maps out your interests—your preferences when it comes to work and hobbies—and your preferred style of approaching projects or decisions.

You get a ton of data, so users typically work through the results with a trained facilitator.

*How this is (and isn't) useful.* I took this assessment along with my entire research group, and many of the students and postdocs said the most useful part of the survey was the section on their interests, as that helped them think about their future career options. I found that the most useful part for me as a leader was learning about the distinctions among usual behaviors, needs, and stress behaviors. This has been helpful in two ways: (1) Conflict is often a result of people not behaving the way we expect them to, and being able to recognize that someone is acting in an unexpected way because they are in their stress-behavior mode can help to prevent or clear up misunderstandings; (2) We assume that if our usual behavior is similar to that of someone else, we shouldn't have conflict, but what is more important is the alignment of our usual behavior with their needs. Take my earlier example of physical energy. If I'm working with someone who has a profile similar to mine, I might assume that because we are both high-energy people, we can drive each other to meet deadlines. However, as we do so, we trigger each other's stress behaviors instead. Simply put, this assessment highlights the fact that the way we act isn't necessarily the way we prefer others to interact with us. I've also found that this assessment is one of the best for discovering unique strengths or characteristics. When reviewing my report, one of my mentors explained, "This metric is emotional energy. Everyone feels a lot of feelings, but this is how much you like to talk about your feelings. Jen, you're a 99—*I'm guessing that's not typical for scientists.*" While I literally laughed out loud at this statement, it did point to something that has deeply affected my career path—I'm a chemistry professor who likes to talk about feelings and emotions. That "weird" combination has motivated my activity on social media, the presentations I give, and even my decision to write this book. Okay, so I clearly love this assessment method. However, as with people, its strength can also be its weakness—users receive so much complex data that it can be a challenge to unpack it all and figure out how to use it.

## How to Choose

Whew. That was a lot. Given so many options, how do you decide which assessment tool to use? There probably aren't any wrong choices, but some options might be more useful to you than others. Among the questions to consider: How much information do you want? Are you willing to spend money on this? Is there an assessment that the people you work with already use? You might also find value in exploring multiple options and then focusing on the one that resonates best with you. Again, the goal is not to capture all of the complexity of who you are, but to give you a framework for thinking about your unique strengths and how you might be able to best use them in leading yourself and others.

## Strengths in Action

You've taken the test, the results are in, you're figuring out what makes you weird. Now what? I've just shared a few examples of how strengths can manifest in your leadership and how this can work for or against you, and psychologist Daniel Goleman offers a much broader view with his description of leadership styles. Based on research from the consulting firm Hay/McBer, Goleman describes six leadership styles that are commonly found in the workplace.[10] While some of these styles have been shown to be more effective than others, they each have contexts in which they are useful, and they each draw on different strengths.

- *Coercive* leadership involves issuing direct orders while asking few (or no) questions and demanding compliance from those who are being led.
- *Authoritative* leadership offers a strong vision for everyone to align with, but does so with enthusiasm rather than fear and empowers people to take ownership in achieving goals.
- *Pacesetting* leadership is, in contrast, very hands-off, in that individuals are given a high level of autonomy and are expected to know how to meet high performance standards.
- *Affiliative* leadership focuses on people rather than productivity, with the attitude that building trust and relationships will translate into motivation to perform.
- *Democratic* leadership seeks to survey the opinions of everyone who is involved and build consensus around key decisions and processes.
- *Coaching* leadership works to develop each person and help them perform at their best, which in turn can translate into positive results for both productivity and personal growth.

Of these approaches, researchers found that the coercive and pacesetting styles were least effective in the workplace. This is especially troubling because these two styles seem to show up frequently in academia. Early in my career I leaned heavily on the pacesetting style—I am fiercely independent and don't want to be given instructions, and so I assumed that the members of my group would all want the same thing. Not necessarily. Leading in this way frequently resulted in confusion on their part about what was expected, and frustration on my part when my expectations were not met. Sound familiar? Many of us are in this job because we prize our independence, and so we may lean heavily on this leadership style. However, we need to recognize that what works for us will not necessarily work for those in our research group.

In contrast, the authoritative style was found to be the most effective, at least in the context of business. Compared with the two styles we just talked about, coercive and pacesetting, authoritative leaders set clear goals and then seek to achieve those through encouragement rather than fear. The affiliative and democratic styles are also effective, though Goleman notes that while they make people feel valued, this can come at a cost to productivity. Interestingly, the coaching style of leadership was found to be used the least often, even though it is also an effective approach. In academia, where my goal is not only to get research done but also to train and equip students and postdocs for their future careers, the coaching style is one that I know I should draw from frequently.

While it is clear that some leadership styles are better than others, Goleman points out that, similar to a set of golf clubs, they each serve a specific purpose. Just as a 3 wood is great from the fairway but a liability on the putting green, using an authoritative style can be highly effective when trying to get data for a grant submission but not when talking with a student who is experiencing a family emergency. Learning to play golf includes developing insight for which club to use in each situation on a course—tee box, fairway, sand trap, putting green—and the skill to use each of the clubs effectively. Similarly, we must become adept with each of the leadership styles so we'll be prepared to use the right one when the situation warrants. That doesn't mean that they will all be equally easy to master. You may find that the goal setting and enthusiasm needed to be an authoritative leader come naturally to you, but developing the patience and listening skills to be a democratic leader is significantly more challenging, or vice versa.

## LEADERSHIP STRENGTHS AND STYLES

As you learn about your strengths, think about how they translate into your preferences among the leadership styles. Since we receive so little formal training in leadership, it's possible that the styles you use most often aren't even based on your strengths but come from emulating mentors or managers you've had in the past. If you've played golf, you probably know that just because someone else is using a 7 iron from the middle of the fairway doesn't mean that you should. When you recognize your own strengths, you can gain a better understanding of which styles best allow you to use them, seek out opportunities in which that style is effective, and learn to lean on the people around you when you need to use a style that falls outside of your natural strengths or comfort zone. We're all a little bit weird, and that's what makes this fun.

## ACTION ITEMS

- *Learn about yourself.* Try using one or more different personality assessments and see what you think. Which one makes the most sense to you? Which one most clearly points to the characteristics that make you who you are? If you're unsure, share the results with a few people you trust; they will likely be able to quickly tell you which test got it most right. If financial resources are limited, you might want to start with one of the free versions of an assessment, or your institution may offer a leadership seminar or short course that includes one of these assessments.
- *Look for opportunities.* Based on your assessment, choose one strength or characteristic that reflects something you enjoy or are passionate about but you are underutilizing in your work. Find an initiative or program that you could get involved in, or

think of something that you could start doing with your research group that would better capitalize on that strength.
- *Excavate what makes you "weird."* While the people around us have a good perspective and can often recognize and articulate what makes us unique, we can also do this for ourselves through self-reflection. Look back at the priority grid that you made in Chapter 1 and pay specific attention to the activities that you put in the "exceptional" category. Do you notice any themes? You can also reflect on how you approach those tasks and how your approach differs from the "typical" way of doing things. Formulate a working hypothesis about your unique strengths, and then continue to test and evolve that hypothesis as you lean into those strengths in your leadership.

# 3

# GOALS

## Where Do *You* Want to Go?

WHAT DO YOU WANT TO ACHIEVE IN THE NEXT FIVE YEARS? As you consider this question, work and personal goals may immediately come to mind. Or the question may give you pause as you realize you don't know the answer. For many faculty, the first five or so years of their independent career are focused on getting tenure. Then, for the next several years they have an almost equal certainty about a different goal—getting promoted. While this may strike you as either obvious or a bit humorous, it also reflects a serious challenge in academia. Our academic system is very good at making some goals loom so large that they overshadow anything else on the horizon. After all, how many industries tell their employees, "I have a goal for you. You have five or so years to achieve it, and then you'll either have a job for life or you will need to find work somewhere else." This creates a culture that pushes us to place a singular focus on that one goal at the expense of everything else. As a result, there's a good chance that you know someone (or you are someone) who made it to tenure or promotion and then fell into an existential quandary as they confronted the question, "Now what?"

Whether or not you've personally experienced this pressure around the tenure system, it highlights the truth that no matter what we do, we all have goals. The real question is whether they are the goals that matter most to us. If we aren't intentional in deciding where we want to go, then the default is to adopt whatever is dictated by the people or culture around us. This sets us up to feel productive as we work toward that goal, but it can leave us hanging with the "now what?" when we discover that success feels hollow when the goal we've achieved was not our own. Even after we achieve key career milestones, it may still feel like someone else is dictating our goals, or that we have little control over our target. After all, we know that we need to get grants, publish papers, teach well, and survive our committee work. But shouldn't there be more to it than that? I hope so.

One way to think about setting goals is to use the analogy of driving somewhere. The first thing you need to decide is where you want to go and why. Letting the academic system set our goals for us is a bit like telling your friend that you are going to drive to the grocery store, and when they ask why, answering, "Well, it seems like everyone else is going there." While this example is laughable, we can let ourselves do the equivalent with our careers if we're not intentional about our goals. In the driving example, once we decide where *we* want to go, we can then plug the address into a GPS app on our phone, and that can provide the possible routes and tell us approximately how long it will take to get there. We choose our route, figure out what direction we need to head in, and we're off. Choosing and mapping out career goals is obviously more complicated than planning a quick drive to the grocery store or even a longer trip, but we can follow the same process, and in doing so, get ourselves to somewhere we actually want to be.

## Defining Your Win

We are innately driven to win, and there's nothing wrong with that being our overarching goal. What matters is how we define "winning." In some arenas of life, the definition is obvious. If you play or enjoy watching competitive sports, you know that winning is often defined by scoring the most points or having the fastest time. I would argue, however, that even in those cases it's not so simple. Think about how we react when we see a baseball player score the winning run in a game and then later find out that they were using performance-enhancing drugs. Or if you were watching a marathon and saw one of the runners hop on a motor scooter for part of the route, you would not consider them the winner, even if they crossed the finish line first. The definition of winning can also vary depending on individual goals. As an amateur distance runner, I am unlikely to win or even place in my age group for any of the marathons I enter. But for each race, I still have a goal time in mind for myself, based on the course and my current level of fitness. If I'm able to cross the finish line in under that amount of time without breaking any rules (or getting injured), then I'm going to consider that a win.

The same concepts apply to our career goals. There are certainly some wins that seem fairly obvious, like getting a manuscript published or a new research grant funded. We also recognize that these are victories only if they are achieved in an ethical manner—in this case, meaning the research in the manuscript was done with a high level of rigor and the results reported with integrity. But what about the less obvious wins? What does it look like to achieve our goals as mentors or leaders? And how do we want to balance those goals with the ones that we have for ourselves outside of our careers?

In his classic book on self-leadership, *The 7 Habits of Highly Effective People*, author Stephen Covey outlines an exercise that can help us push through the noise of the people and culture we're surrounded by and articulate goals that are unique and meaningful to us as individuals. His exercise is based on imagining yourself at your own funeral, but I personally prefer to think about my own retirement party. Either way, the instructions are to visualize yourself listening to a variety of people talk about the role you've played in their lives and think about what you would want them to say. In an academic version of the exercise, this may include members of your research group, students you've taught in classes, your colleagues and collaborators, your dean or department chair, and finally your family and friends. It definitely takes some effort to fast-forward a few decades and imagine what you would want to be remembered for, but what you come up with paints a picture of your definition of winning.

## Mapping Your Route

We're going to keep imagining, but this one is a bit easier because you only have to look five years into the future. Similar to what we discussed in Chapter 1 about getting in touch with your future self, take a moment to think about where you will be in five years—your age, your family and friends, and your career stage. Now think about what you want that future to look like. Do you hope to have expanded your research program into a new area? How big do you want your research group to be, and how much funding will it take to sustain that? What new topics or strategies do you envision incorporating into your teaching and mentoring? Are there any specific leadership or service roles that you are interested in? If you get stuck, go back to your priority table from Chapter 1 and look at the tasks that you put in the "exceptional" column—these are the things that you enjoy doing and are willing to put your time toward, even though they are not critical job duties. Finally, as you look over your list, ask yourself whether your five-year goals are in line with where you want to be at the end of your career, and adjust as needed.

Congratulations! You have a pretty good idea of where you want to go. Now it's time to get out the GPS and map your route. When driving, if you are heading somewhere far away and unfamiliar, you know that your likelihood of reaching your destination will be low if you simply glance at the map one time and then turn off your phone. The same is true with goals, as research shows that we are less likely to achieve goals when they are complex.[1] Just as we can navigate a complex path in real life by having a list of directions and receiving continuous feedback, we can make complex goals more attainable by breaking them up into smaller tasks and regularly tracking our progress. For each of your big five-year goals, ask yourself what you would need to do in the next year to work toward it. Then think about the next month and the next week.

Let's work through an example together. It will come as little surprise to you that a few years back, one of my five-year goals was to write a book. While I was excited about the goal, it felt exceptionally intimidating. So, I set a one-year goal to create an outline of the sections and chapters. I also recognized that I had no idea how to go about publishing a book, and that it would be good to get that figured out before I started actually writing. So, for the first month I set goals of spending at least two hours reading articles online about how to publish a book and creating an outline that included at least three ideas for chapters. My one-week goal was to open a Word document on my computer and start my outline by just writing something—anything.

No matter what your ultimate goals are, there are a few general principles you can follow in creating and working toward all of your intermediate goals along the way. First, make sure they all line up. Your one-week goals should work toward achieving your one-month goals, which should work toward achieving your one-year goals, and those should represent reasonable steps toward your ultimate five-year goals. This may sound obvious, but applying this principle serves as a great check on your backward design process, and I usually do find places where I can modify goals to achieve better alignment.

Second, set "SMART" goals. First introduced by George Doran, the SMART mnemonic has become a widely used guide for effective goal setting and is backed by studies in goal theory.[2] While there are many variations on the mnemonic, the following is one that I find works well for academic goals:

- *Specific.* The goal clearly describes what you want to achieve.
- *Measurable.* Outcomes can be quantified (time spent, words written, applications submitted).
- *Attainable.* You can realistically achieve the goal given your time and other available resources.
- *Relevant.* The goal is part of a bigger plan and is focused on something that is important to you.
- *Time-based.* There is a set deadline for completion.

Looking back at my example, the goals I outlined adhere pretty well to the SMART criteria. I will admit that this isn't the case for all of the goals I set, but the important thing is to come as close as you can to following these principles. Additionally, create a schedule for your one-week and one-month goals, including both when you will meet them and a plan for maintaining progress even if you encounter a challenge. Generating a schedule and a backup plan creates what is called an implementation intention, a strategy that has been shown to increase your likelihood of success by lowering the barrier to getting started and reducing the chances that you will be derailed by an obstacle or failure.[3] Much like planning to drive somewhere, you can sit and stare at your map all you want, but if you are going to reach your destination, you need to start the car and make the first turn out of your driveway. As a side note, using the SMART framework and implementation intentions can also be highly valuable when crafting aims for grant applications, running a lab, or mentoring students through the process of project planning, so whatever practice you get here with your own goals will pay dividends in other areas of your scholarship and leadership.

The third principle is to share your goals with others, as this can also increase your likelihood of success.[4] While this may sound unpleasant or awkward, it doesn't have to be. You don't have to sit down with someone and say, "I'd like to share my goals with you," and then pull up a Word document on your computer. As an example, I shared my goal of writing a book during a New Year's Eve dinner with my spouse, John. In reading through my story thus far, you might assume that I set a goal to write a book, decided to go for it, and then dove right in. Not so. In reality, I had been tossing the goal around in my head for several months, battling between my enthusiasm for putting my ideas out into the world and my self-doubt, which was telling me that I could never make it through such a huge project. In the midst of this, John and I happened to be out for dinner and our conversation turned to our individual goals. After a glass of wine, I cautiously admitted that people had been telling me that I should write a book, and that while I had some *significant* doubts, I was at least considering it. His response was so perfect that I'll never forget it. He looked at me and said, "Well, what would it look like to try?" And here we are. Sharing your goals with someone you trust can certainly be beneficial in terms of creating accountability to ensure that you follow through on them, but perhaps even more important, it can give you the nudge you need to take a step forward, and let you know that there's someone in your corner cheering you on.

Finally, as you work toward your goals, create a schedule for checking your progress and outlining next steps. Depending on how much you like structure and planning ahead, you may prefer to go with my approach of setting your intermediate goals one week or one month at a time and frequently coming back to update them. Or you may prefer to look at your one-year goal and outline in advance what you will do during each of the fifty-two weeks between now and then. Either approach is fine, and this is a great opportunity to lean into your strengths. What's important is that, however you structure them, your long-term goals become shorter-term tasks that then become items on your to-do list or schedule, so that you can track your progress.

It's also important to recognize that five years is a relatively long time and your goals may change over this period. Sometimes a change is catalyzed by an external event, such as being nominated for an advisory board position or being approached for a new collaboration. In these instances, having your five-year plan can help you evaluate whether the new opportunity is one you are interested in. In my case, when I received an email asking if I would consider applying for the department chair position that I currently hold, the opportunity represented a significant pivot in my path, as it involved a cross-country move and a new set of duties. Still, the decision to apply for the job was an easy one to make. Pursuing an academic leadership role was one of the five-year goals I had outlined, and the department's vision for its future aligned very closely with my ultimate "win" of creating a healthier academic culture. Sometimes the unexpected news that prompts a pivot comes in the form of a failure or rejection, and I'll share more about how to manage those instances when we get to Chapter 5. In addition to evaluating opportunities and managing failures that arise unexpectedly, you can also engineer pivots into your five-year plan. Just as your GPS app regularly checks to make sure the route you're on is still the best option, you can make it part of an annual or semiannual check-in to reevaluate your plan and modify as needed.

## Be Careful What You Catch

Even if you don't sense it, the people you surround yourself with have a significant impact on your goals. At the start of this chapter, I talked about the tendency to fall into the default goals of the academic system, but there is an even more specific impact from the individual people that we surround ourselves with on a daily basis. Psychologists who study the ways that people set and achieve goals observe a principle they call "goal contagion."[5] This is exactly what it sounds like. Just as someone can infect us with their cold by sneezing in our coffee cup, the people around us infect us with their goals—both good and bad—through our interactions in hallway conversations and department happy hours. Interestingly, psychologist Chris Loersch and co-workers have found that goal contagion is strongest when people identify with the same group.[6] This means that if you see yourself as a faculty member or an academic, you are most likely to "catch" the same goals of the faculty and academics you spend time with.

Just as with avoiding a cold, one way to protect yourself from this is to take a careful look at the health of the people around you and adjust accordingly. Now that you've outlined where you want to be in five years and at the end of your career, take note of the goals that are being projected by the colleagues that you spend the most time with. Even if you never sit down and ask them, "What are your goals for the next five years?" you probably have a good idea of what they are just from what each person says and does. If you're fortunate, you already have some friends and colleagues who have healthy goals that align well with yours. On the other hand, this can be challenging if your department or social circle is relatively small or if people are focused on metrics that hold little interest for you. You may also have some goals that are not typical among the people in your field (say, for example, you're a chemistry professor who wants to write a book on leadership). In either case, social media and other networking opportunities can be very helpful for finding people who will reinforce your goals. One key thing to remember is that this is a mutual process. When you do identify that group of people and intentionally spend time together, recognize that while you are benefiting from the goals that they project, you also have the potential to encourage them in their goals through what you project back.

## The Comparison Game

While it's important to be mindful of who you're surrounded by, paying *too* close attention to other people's goals can be detrimental to your success and happiness. This often comes in the form of comparison. A famous quote widely attributed to Teddy Roosevelt is, "Comparison is the thief of joy." At the very least, it can be the thief of time. As we discussed in Chapter 1, most of us feel like we don't have enough time to get everything done as it is, and spending hours obsessively tracking the publication records of our peers and comparing them with our own is not going to help (trust me, I've been there). But not all comparison is bad. In some cases, it can provide useful information about the metrics of success. So, how do we get the information we need without derailing our happiness or our schedule?

One challenge lies in recognizing that we tend to distort reality when we make comparisons. Not only do we compare ourselves with the very top performers, but we do this separately for each dimension of achievement.[7] For example, you may be simultaneously comparing your teaching evaluations with those of the best instructor in your department, your publication record with that of the most active researcher in your field, and your mentorship with that of the most experienced colleague you know. In addition to creating unrealistic standards in each individual area, together these comparisons generate a picture of success that is completely unattainable. Those individuals who excel in one area often do so at the expense of other areas, for example by devoting less time to teaching in order to write more papers, or submitting fewer grants in order to have more time for hobbies outside of work. If we are going to compare ourselves with others, we have to compare a whole picture with a whole picture rather than comparing our whole picture with a mosaic of highlights from others. This is also where knowing your "win" is incredibly helpful. You might still feel a twinge of envy when you see someone receive an award or be recognized for a specific achievement, but if that particular activity is not in line with what you ultimately want to be known for, it is much easier to banish the jealous feelings and instead experience genuine happiness for them.

Your specific short-term goals can also allow you to look to others for inspiration while staying focused on your own metrics of success. As a recreational rock climber, I enjoy watching videos of professional climbers tackling extremely challenging routes. I could be tempted to envy their success or feel frustrated because I couldn't make even the first move on the routes that they seem to float up effortlessly. Neither reaction would be healthy, though. Instead, I enjoy watching the videos to get inspired by what is possible, and then I apply that inspiration to my own goal of climbing a route that is at my limit. I'll admit that in this example it's easy for me to maintain a healthy attitude—after all, most professional climbers are decades younger than I am, and they spend all of their time climbing, whereas I spend my days mostly in front of a computer screen. But the principle can still apply in my work life. Even when we are in a similar position and at a similar career stage as someone else, we each have different priorities and work in different places. That means we should each have our own set of goals, and the better you can articulate yours, the more you can stay focused on achieving them and enjoy a sense of satisfaction when you succeed.

Finally, as you do plan out your goals, you may realize that many (if not most) of them are not based on a zero-sum game. My goal to write a book is not impeded by someone else doing the same. Your pursuit of a new research area is not going to be prevented by your colleague's decision to expand their research program. We need to remind ourselves that someone else's success does not have to detract from our own. And even when we do find ourselves in a situation, such as competing for an award, when one person's win means another person's loss, there is still a strategic benefit to putting aside competitive jealousy and working together with others. Think back to when you applied for jobs. There are always a limited number of openings, and when one person receives an offer it typically means that someone else doesn't. Nevertheless, there is a good chance that you still learned something from fellow applicants and benefited from your support network. Did you get examples of applications from others? Share information about new job openings? Read over each other's teaching and research statements and offer feedback? Each of these actions has the potential to increase your success and, when taken among a small group of friends, costs you very little. You may have also experienced the benefits of collaboration in your research. Although you are competing against other researchers for grant funding, working together can allow you to tackle more ambitious projects and publish more impactful work, in turn helping everyone be more successful in securing funding. The underlying principle here is that even when things are a competition, collaborating with a smaller group of close colleagues can provide important support and help each of you to be more successful in achieving your goals.

## ACTION ITEMS

- *Plan your retirement party.* Schedule some time to work through the retirement party exercise. You can do this all at once in an hour or two, or you can tackle one group of people each day over a week. There aren't really rules here, so you can write out quotes or just make bullet points. Once you're done with that, take a look over everything that you want said about you. These statements will reflect your values and what you consider "winning." Pull out three to five themes that you can hold onto and use in setting your goals.
- *Set some SMART goals.* If you haven't already, map out your goals by working through the five-year, one-year, one-month, and one-week goals (or whatever time frames work for you personally). Figure out your first steps and put those on your calendar or to-do list. Share your goals with someone who cares about you. Doing so over a glass of wine is optional.

- *Survey your network.* Make a list of the people you spend the most time with. Think about each one and the extent to which they share your goals and your approach to success. Look for gaps. Do you have a big goal that nobody close to you shares? If this is the case, look more broadly across your network, or even expand your network using social media, to find people you can spend time with who will reinforce your goals.

# 4

# MOTIVATION

## What to Do When You Don't Want to Do Anything

I'M STARING AT A BLANK SCREEN, hoping that somehow the words I need to write will materialize in front of me. I find this project daunting and I'm tired, even though I normally love writing. Maybe I'll take a little break to help me get motivated, perhaps check my email . . . and then social media . . . and then get another cup of coffee . . .

It's June 2021 and I'm sitting in a cabin in the North Georgia mountains. We are over a year into the COVID-19 pandemic, and I've sent myself on a short writing retreat to jump-start my work on this book again. I started writing at the beginning of 2020 with a goal of completing one chapter each month. I made it through three chapters, and then March 2020 hit. Seemingly overnight, our world changed dramatically, and my spouse and I found ourselves trying to balance homeschooling two kids while still doing our "day jobs" of teaching classes, serving on committees, and managing a lab. Almost everything that wasn't on the "essential" list came to a halt.

This feels like the worst possible time for me to write about motivation, but maybe that also makes it the best time to do so. Rather than feeling energized and hopeful, I feel burned out and discouraged. Whatever the reason, maybe you feel this way too. But we don't have to stay in that place. The principles and practices we're going to explore in this chapter can provide the tools we need to rebuild our motivation, whether we're living through a global crisis or struggling to balance career and elder-care responsibilities or simply trying to manage the daily procrastination cycle. Just thinking about this makes me feel a little more motivated.

## Procrastination Station

Procrastination station is not a real place, but I somehow still visit it often. And I spend enough time on social media (while—ahem—procrastinating) to know that I'm not alone. The struggle with motivation is not unique to academia, but it certainly seems to be amplified here. If you've earned (or are working toward) a PhD, you are probably someone who has experienced some level of success academically and enjoys seeking out new challenges. While it may seem counterintuitive, this actually makes you the perfect candidate for procrastination. Research from psychologist Joseph Ferrari and coworkers has shown that procrastination is higher among college students with a history of strong academic performance compared with those who perform at average levels, and that procrastination increases as students advance through their college careers and become more self-directed.[1] According to this trajectory, by the career stage at which you are reading this book, I'll bet you are an expert procrastinator. Instead of feeling down about this, just keep reading—you may come to appreciate how the right type of procrastination can actually be a strategy for success and effective self-leadership.

In his book *Solving the Procrastination Puzzle*, psychologist Tim Pychyl outlines the types of tasks that we tend to procrastinate about. His list includes tasks that are difficult, frustrating, ambiguous, or unstructured. You may not use those exact words, but nearly every one of the tasks that you listed for your time-management grid in Chapter 1 probably fits into one of those categories. As an example, let's look at writing a grant proposal. It is definitely difficult, as only 10 to 20 percent of proposals are likely to be funded. There will be frustration every time you encounter a potential flaw or weakness in your research plan. It is ambiguous, in that no one has ever written that exact proposal before. And the entire task is an unstructured project that you can tackle in multiple ways over anywhere from many days to several months. On top of all of that, the nature of an academic job means that you don't have a boss, and unless you are working with collaborators, nobody is checking in on your progress on a regular basis. No wonder we are so quick to pick up our phone and check our email in the secret hope that we will see something that requires an "urgent" response and thus give us license to set aside the grant writing until later.

Ironically, procrastination can also be a key to our motivation and creativity. In his groundbreaking 1990 book *Flow: The Psychology of Optimal Experience*, psychologist Mihaly Csikszentmihalyi describes the experience of being so deeply immersed in an activity that we tune out the world around us and become completely absorbed in the task at hand. Also referred to as being "in the zone," achieving a flow state floods us with confidence and self-efficacy because we feel that we are performing at the top of our game. What is surprising is that one way we achieve flow is through procrastination. In an analysis of procrastination in college students, researchers found that students often use procrastination to create a "peak affective experience" in which they have a tight deadline for a challenging task, and this pushes them to eliminate distractions and perform at their best.[2] In other words, while procrastination can be detrimental to our progress in the short term, it can also create the ideal conditions for us to enter into flow and, as a result, find both motivation to perform at our best and a deep sense of satisfaction once the task is complete.

So, how do we ensure that our procrastination navigates us toward a state of flow rather than to a missed deadline because we were too busy watching funny cat videos? Similar to competitiveness, which we saw in Chapter 3 can be either helpful or harmful, procrastination can be either active or passive, as highlighted by organizational psychology researchers Angela Hsin Chun Chu and Jin Nam Choi.[3] When we actively procrastinate, we intentionally wait to start a task so that the amount of time until the deadline will create pressure and, in turn, push us toward flow. In contrast, when we passively procrastinate, we simply put off doing something even though we know it needs to get done. The difference between these is our level of control—active procrastination is healthy because we are exercising control and choosing to delay starting (and hopefully choosing other productive things to do in the meantime), whereas passive procrastination represents an inability to control our choices. We'll circle back to flow in a little bit when we consider how to manage motivation. First, we need to dive into a deeper question.

## Why Are You Here?

Given what you've just read about procrastination, I need to provide a disclaimer with this section: you're about to tackle something that may be difficult, frustrating, ambiguous, and unstructured—figuring out why you chose the career path that you did. As you'll see in a moment, I'm not talking about the easy answers like "I love research." Rather, I'm inviting you to think deeply about what fundamentally motivates you. Getting at the real answer will take some time, but I promise it will be worth it.

Many of my hobbies are athletic, and I often read articles or listen to podcasts that deal with athletic training or sports psychology. A few years ago, I read an interview with a personal trainer that stuck with me. In the interview, the trainer described the first conversation they have with each new client who is thinking of signing up for a training program, and how they ask the seemingly simple question, "Why do you want to do this?" The trainer said nearly every client offers one of two general answers: (1) They want to lose weight and/or feel more attractive, or (2) They want to be healthier. In my rush to judgment, I assumed that the first answer was incorrect and the second was correct. Thus, I was surprised when the trainer explained that both answers are incorrect—or at least insufficient. "I want to be healthier" might work during week one of the program (or on January 2, after a New Year's resolution), but it will fall short when it's week eight, the client is feeling tired, and it's snowing outside. Getting motivated to get out the door on those days takes a much more specific answer. The trainer then described how they just keep asking "why" and probing with questions until they reach the real answer. I don't remember the exact examples from the interview, but the real answer could be something like "My dad had a heart attack at age forty-two and I'm about to turn forty. I don't want my kids to go through the pain and worry that I did," or "I know that I'm lonely because I don't have enough confidence to meet new people. If I start working out, that will give me the confidence I need to form new relationships." An answer with this level of detail is much more likely to compel someone to pull themselves out of bed and into the snow for a cold morning workout in February.

You can apply this same principle to finding motivation to get through the tough days at work—those times when the to-do list feels overwhelming, your calendar is stacked with meetings, or you need to revise and resubmit that manuscript for the third time. Knowing the real answer to why you chose your current career path can also be a very powerful guide when making future choices—once you know what you really love about your current position or responsibilities, you can look at each future opportunity and ask, "Will this give me more or less of that?"

So, how do you find your career "why"? Like the personal trainer, you have to keep asking questions. If you ask me why I love what I do, my gut reaction is to say it's because I love science. That's not incorrect, but it's the equivalent of saying I want to exercise to be healthier. There are tons of jobs that involve science, and just "loving science" probably won't get me out of bed in the morning and motivate me to work at my top capacity. If asked why I love doing science, I would tell you (with some help from my Birkman analysis) that it's because I'm very independent, and I thrive on the autonomy and creativity provided by research. That's a better answer, although it still isn't specific enough to explain the career choices I've made. Pushed further, I would say that I also find joy in building relationships. This is yet another step closer, as it gets to a specific part of my job and helps explain why I approach the job the way that I do. But there are lots of jobs that involve science, independent research, and building relationships. After taking a few more steps down the road, the answer I finally arrived at was: I love working with an ever-evolving group of energetic, creative, and kind students and postdocs, getting to see them grow in their knowledge and independence during their time in our group, and watching their success as they move on and navigate their future career paths. Now *that* is a "why" that I can draw on for motivation. As you might guess, one of the action items at the end of this chapter is going to be identifying *your* why. Feel free to start thinking about this now or to engage in active procrastination until you get there.

## Going to the Source

The examples we just discussed highlight why understanding the basic psychology of motivation can be so important. If our motivation were never broken, we would never need to fix it, but that isn't how life works. We will all have times (maybe many times a day) when we don't feel motivated to do something we need to do, and understanding the sources of motivation can be incredibly helpful in getting ourselves through those moments.

One practical framework for thinking about this topic is that of intrinsic versus extrinsic motivation. Popularized by the research of psychologists Richard Ryan and Edward Deci, this framework identifies two potential sources of motivation:

- *Intrinsic motivation* drives us to do something because we find it inherently fun, interesting, or enjoyable.
- *Extrinsic motivation* drives us to do something to generate an external outcome, such as attaining a reward or pleasing another person.[4]

Thinking back through your life, you can probably identify many times when you have been driven by each of these types of motivation. If you have a hobby that is unrelated to your profession, there is a good chance you are intrinsically motivated to pursue that activity. And if you ever raced home as a teenager to make it before curfew, then you're very familiar with extrinsic motivation. Similarly, the tasks you do at work will generally fall into these two categories. On the intrinsic-motivation side, you might read a journal article solely because you find it interesting, or start a new collaboration because the research sounds fun and you are excited to learn about a new discipline. There is also no shortage of tasks that are extrinsically motivated, such as course grades that need to be finalized by the university deadline, progress reports that need to be submitted to maintain funding, and manuscript revisions that need to be completed to add a publication to your CV.

You may think that intrinsically motivated activities are less important to your work because they are voluntary. However, research shows that we are more creative and do better work when we are intrinsically, rather than extrinsically, motivated.[5] Additionally, studies have shown that, whether in academia or sports, the source of our motivation can have a significant impact on burnout, with intrinsic motivation serving as a protective buffer.[6] Finally, thinking back to our earlier discussion of flow—that sense of being in the zone, where we're having fun being immersed in a challenge and deriving a high level of satisfaction from our work—it probably comes as little surprise that it is closely linked to intrinsic motivation.[7]

A very real challenge is that many of our job responsibilities are extrinsically motivated tasks, and that sets us up for nonideal performance, less enjoyment, and an increased likelihood of burnout. This is where the self-leadership part comes in. We may not be able to control the things that we need to get done, but we can control how we think about those tasks and how we manage our motivation in pursuing them. While it may sound like we're just fooling ourselves by doing this, convincing our brain that something on the have-to-do list is actually a want-to-do activity can move it from the extrinsic to intrinsic column and provide the associated benefits. Granted, it takes some creativity and self-awareness to be able to fully reframe to the point of saying, "I just can't wait to complete that grant progress report!" but it can be done, and it can have significant benefits.

## Three Motivational Movers

Ryan and Deci highlight that there are actually multiple forms of extrinsic motivation, some of which provide benefits similar to those found with intrinsic motivation. They identify three factors that allow us to move along the motivational spectrum: *regulation, competence,* and *relatedness*.[8] In his book *Drive: The Surprising Truth about What Motivates Us,* Dan Pink expands on this idea with the terms *autonomy, mastery,* and *purpose*.[9] Pink proposes that we can tap into intrinsic (or intrinsic-like) motivation by finding or creating these factors in our work. Even if you're not familiar with these terms in the context of motivation, you have probably experienced their effects. Think, for example, about the times that you have had the most fun or felt most fulfilled in your work. They likely involved having the freedom to be creative, gaining new knowledge and honing your skills, or making a positive difference for others—that is, they gave you a sense of autonomy, mastery, and purpose.

How do we apply this in everyday life? One exercise I like to use focuses on choosing a task that I'm not looking forward to (or I am procrastinating about) and then identifying in it the threats to and opportunities for autonomy, mastery, and purpose. This helps me to see why I'm not feeling motivated and how I can move in the direction of intrinsic motivation. As an example, we'll look at something that I should love doing but secretly kind of dread: reading journal articles. If I'm candid with myself about why I dread this, I can identify threats to my motivation in all three areas.

- *Autonomy.* I feel like I need to look through all of the articles that show up in my weekly keyword search reports, so I have very little freedom in this task.
- *Mastery.* When I read an article about a really clever idea or research I don't fully understand, it makes me feel unintelligent. And if I don't retain everything I read, I feel like I'm losing knowledge.
- *Purpose.* When I check the literature, there is always the risk that I will learn that someone has "scooped" our lab, and that will have a negative impact on the researchers in my group.

While these thoughts are real, you might notice that I also have some false perceptions mixed in. If I read something and then forget part of it later, I won't actually lose knowledge—I will still know more than if I'd never read the article in the first place. And if our lab gets scooped, that will happen whether or not I actually read about it in a publication.

When I reframe those initial threats as opportunities, I arrive at new sources of motivation.

- *Autonomy.* I do need to check my keyword searches and read the publications relevant to my research, but I can also give myself the freedom to browse through a set number of articles each week that are outside of that list and "purely for fun."
- *Mastery.* Even if I don't fully understand a paper or later forget part of what I read, I will still be learning something that makes future papers on that topic easier to read, and I'll be creating an opportunity to move my research in a new direction.

- *Purpose.* The more I read, the more ideas and techniques I'll be able to use to help the members of my group succeed in their projects.

In thinking about that last bullet point, you can probably envision a number of ways to view reading journal articles as having purpose. For example, reading the literature can equip us to do research that benefits society—by addressing climate change, for example, or improving human health—and that is an important purpose. This is another place where your answer to the "Why are you here?" question is important. Reframing a task as an opportunity for autonomy, mastery, and purpose will be most effective when your reasoning aligns with the fundamental thing that drives you in your work. My source of purpose in reading the literature originates in helping the members of my group succeed, as that is what motivates me. On the other hand, if you are primarily driven by wanting to invent a new drug to cure a disease, then you would write something very different in your "purpose" bullet point. There are no right or wrong answers, only those that work best for you.

I've found that this exercise is helpful for almost any task. By harnessing some of this intrinsic-motivation creativity, you can make even the most unpleasant task a bit more bearable or make mundane tasks sound fun and interesting. And the more you practice this reframing, the more ingrained these thought patterns become.

## Other Factors at Play

When it comes to managing our motivation, I've found that the internal reframing and dialogue I just described have by far the greatest impact. This makes sense, as we operate within our own thoughts every waking minute of every day. Still, there are a few other important factors to consider as you master this aspect of self-leadership.

*Motivational contagion.* In Chapter 3, we talked about how goals can be contagious and can spread to you via the people you spend the most time with. It turns out that the same thing has been observed for motivation. In a study of teacher-student relationships, researchers found that even when the teachers didn't know whether they were intrinsically or extrinsically motivated, the students' perception of their teachers led them to mirror their intrinsically or extrinsically motivated behavior. In the case of intrinsic motivation, students became more creative and found greater enjoyment in what they were learning.[10] If you teach, this is certainly something to think about the next time you walk into the classroom. It also brings us back to the question of who you surround yourself with at work. The next time you're in a conversation with a group of colleagues (or perhaps at a faculty meeting), take notice of how people describe the work that needs to be done and what motivations they are communicating through their words. Are people motivated by the opportunity to engage in creative problem solving, learning, and exploring? Or are they motivated by adding lines to their CV, proving that they're better than others, and enjoying their power? Thankfully, in order to benefit from motivation contagion, you don't need to change other people's behavior, you just need to change your beliefs about it. If you can convince yourself that the people around you are in their jobs because they love learning and helping others (this might be a stretch for some of your colleagues, but it's worth a try), it could have a positive impact on your motivation and work as well. This reframing could even spur you into discussions with those colleagues in which your motivations end up having a positive influence on them and, in turn, lead to tangible change that benefits the culture and policies in your department.

*Leading versus lagging measures.* No matter how much we try to reframe our work to focus on intrinsic motivation, the extrinsic rewards and penalties are nearly impossible to ignore. Grant funding, publication records, and teaching evaluations have a significant impact on our career progression, and we need to keep track of them. But we can benefit by focusing on the right metrics—specifically, those that are most closely tied to effort and offer the earliest signs of progress. As an example, our lab used to set our publication goals according to when papers were published. In business speak, that is a "lagging measure" because achievement of the goal is very distant in time from the effort that makes it happen, namely running experiments in the lab and writing the first draft of the manuscript. Recently, we shifted to setting our goals according to when manuscripts are submitted for publication. This can be considered a "leading measure" because it is an earlier indicator of our productivity. Importantly, these two measures are closely tied—we almost never submit a manuscript that is not ultimately published.

Switching our goal from publication to manuscript submission may seem like an insignificant change, but it made a huge difference in our lab's motivation. That is not only because it offered a quicker reward in response to effort, but also because it focused the timeline on something we had the ability to influence. We have significantly more control over how quickly we gather data and write and submit a manuscript, compared with how long it will take a publisher to collect peer reviews or how many rounds of revision will be needed before our work is ultimately published. Not every task has leading and lagging measures, but whenever possible, it is helpful to set goals and create a reward system around the former.

*Burnout is real.* No matter how superhuman you think you are, we all need a break sometimes. Perhaps you've already had the opportunity to establish habits around taking time for self-care, as we talked about in Chapter 1. And as discussed earlier in this chapter, framing your work around autonomy, mastery, and purpose can help insulate you from burnout by allowing you to drive yourself from intrinsic motivation. Still, you may find yourself struggling from exhaustion or a lack of desire to do your work, and it's not your fault. Deadlines, interpersonal challenges, family emergencies, global pandemics, and any number of other factors can push you over the edge, even when you practice healthy habits to manage your time and motivation. You can find information and surveys online or talk with a mental health professional to help diagnose burnout, and the tools we've talked about in this chapter can also be useful. If you notice that you are engaging more than usual in passive procrastination or you are finding it increasingly difficult to reframe your work around intrinsic motivation, these could be signs that you need to take a break or seek professional help. Even if your schedule and responsibilities preclude you from taking a break immediately, find a time in the near future when you can step away and recover (and then put that break on your calendar). Your motivation will be waiting for you when you return.

## ACTION ITEMS

- *Identify your why.* Answering this question for yourself will take some time and mental processing. Start by asking yourself why you show up to work each day, and go from there. If you get stuck, think about the times when you feel most fulfilled by your job, or pull up your priority grid from Chapter 1 and look for clues in the items you identified as "exceptional." If you took one of the assessments discussed in Chapter 2, you may find inspiration there, as well. For examples from other leaders and a

more detailed framework, a resource you may find helpful is *Start with Why*, by leadership expert Simon Sinek.
- ***Optimize your procrastination.*** Open your to-do list and locate the thing that you know you need to do but keep putting off. You know which one. Take a moment to think about how you're procrastinating—is it passive or active? If it's the former, think about how you can reframe the task to find motivation through more autonomy, mastery, or purpose. If the latter, make a plan to procrastinate productively. Figure out when you need to start working on the task so you will feel some pressure but still realistically have time to finish. Then, rather than spending the time between now and then on a movie marathon, think about what else you can get done in the meantime. Perhaps most important, give yourself credit for using procrastination to your advantage, instead of beating yourself up for leaving that task undone.
- ***Measure it.*** Look at the list of goals you have set for yourself or your group (and if you don't yet have a list of goals, now is a great time to make one!). Ask yourself whether the goals are based on leading or lagging measures. For each goal, think about how you can move your measure of achievement closer in time to your input of effort, or how you can quantify success by a metric that you have more control over.

# 5

# RESILIENCE

## Coping with Failure and Success

"NOT DISCUSSED." My brain couldn't even begin to process the words on the computer screen. This was the outcome of my first major external grant application. My lab had collected exciting preliminary data, I had spent almost a month articulating our vision for what we would do with five years of funding, and my peers who read the application had offered encouraging feedback. The text flowed and the impeccably crafted figures were perfectly aligned with the rows of justified type. For four months of waiting during the review process, I had remained hopeful. Now that hope was decimated. Receiving a "not discussed" means that your proposal has no chance of being funded. In fact, it wasn't even good enough to warrant discussion by the review panel. I was crushed. Oh, and did I mention that I needed to get a grant funded in the next few years in order to get tenure? So, I was crushed *and* terrified.

This story highlights one of the least fun parts of an academic research job, especially when the stakes are high: the frequent experience of failure and rejection. In the case of grant applications, the livelihood of our research program (and in some positions, our own livelihood) can be on the line. In addition, the turnaround time for trying to reapply can be on the order of months to more than a year—or there may never be another chance. And grant applications provide just one of many possibilities for failure in academia and in research. There are also the experiments that don't go as planned, manuscripts that are rejected, awards we don't receive, students and postdocs we try to recruit but who join other labs or other programs. The opportunities to be disappointed in this job are endless.

Coping with failure can be especially difficult for academics, because most of us got to where we are by succeeding. Moreover, when we look around at our peers, what we primarily see is all of their successes on display—their lists of publications, funding, awards. Even though we know, intellectually, that funding rates for grant applications tend to hover in the range of 10 to 20 percent, and thus statistics dictate that the people around us are actually failing most of the time, it is easy for us to feel like we are all alone in our failure. That is exactly how I felt. In my mind, I was the only one facing the shame and embarrassment of the "not discussed," while everyone else was receiving good news that day.

## Everybody Fails

Because failure is an integral part of this job, it is absolutely necessary that we learn how to cope with it and keep moving on. Malika

Jeffries-EL, a highly successful chemistry researcher and academic leader, advises that "true champions do fail, they just don't quit when they do. It's not how you start that matters, it's how you finish."[1] In my own experience, I did eventually gain the perspective to realize that I was not as alone in my disappointment as I'd thought. In fact, given the structure of the review panels at that funding agency, literally half of the people who had submitted applications alongside mine had also had their hopes dashed by the result of "not discussed." That knowledge alone took away much of the sting of shame for me. Additionally, of the proposals that were discussed, more than half of those would still not have received a good enough score to be funded. On that day, many more people were feeling disappointed than elated about their chances of funding. When I looked at it this way, the outcome of my proposal review went from being an epic and embarrassing failure to being just a bump in the road. In the decade or so that has passed since that first failure, I've received many more rejections from funding agencies, but the good news is that my lab has also experienced a number of successes, and the same is likely true (or will be true) for you.

One of the more valuable pieces of advice I've received in my career came from a colleague, who told me, "The people who are ultimately successful aren't necessarily those who succeed on the first try, but rather those who just keep trying." While this may seem self-evident, it can be surprisingly difficult to believe when we're faced with a big failure or rejection. This underscores one of the challenging ironies of leadership—the situations in which the success of our team most strongly relies on our ability to show up and persevere are the situations in which it is most difficult for us to do so. You probably know this well if you've had to tell your research group that a proposal has been rejected. This news has real implications for the future of their work, and thus we may feel like it has real implications for our influence as leaders. As researchers, we are constantly judged by the quality of our ideas, and that is an inherently subjective measure. So, admitting that other people have judged our ideas to be unworthy of support can feel like it carries the risk of undermining our team's confidence in our capabilities. After all, who wants to follow a leader who is not capable?

Great leaders know that these are the moments when their words and actions can become a self-fulfilling prophecy. If you believe that a failure means you are not capable of doing your job and signals an unrecoverable outcome, then the people you lead will also believe this, initiating a downward spiral. On the other hand, if you can be candid in sharing your disappointment while remaining confident in your (and their) capabilities, and then craft a vision for the next steps, you can earn even more respect and influence as a leader than you had before. As I'll discuss in Chapter 14 when we talk about vulnerability, people don't necessarily want to be led by someone who never fails. They want to be led by someone who can serve as a mentor and role model for how to confront failure, chart a path forward, and keep driving toward success.

## Confronting Your Fears

To effectively lead through failure, you first must work on confronting your fears surrounding failure. This process starts with taking a close look at why failure is so painful. Oh, goody, right? I doubt you're thinking, "Yes! I'd love to dig in and study that really painful thing up close." But the interesting thing about failure is that the more you understand it, the less it can hurt you.

Motivation researcher David Conroy has studied failure in many contexts and has specifically focused on the factors that drive our fear of failure.[2] He identifies five key categories of this fear:

- Fear of experiencing shame and embarrassment
- Fear of devaluing one's self-estimate
- Fear of losing social influence
- Fear of having an uncertain future
- Fear of upsetting important others

You can see how these fears played out in my situation in the previous section—my fears that our proposal's failure would undermine my own confidence and self-efficacy, that I would lose the respect of my group members, and that it would have real consequences for our lab's future ability to do research. So, how do we manage fear of failure? To answer this question, we need to take a look at how we view our abilities.

Psychologist Carol Dweck has spent decades studying this topic in the context of implicit theories of intelligence, more popularly known as mindset.[3] Her underlying hypothesis is that the assumptions we make about our abilities have far-reaching consequences for how we view effort, challenges, feedback . . . and failure.[4] Dweck outlines two different ways that we can view our abilities:

- *The fixed mindset.* Our skills and talents are unchangeable. We are born good at some things and not so good at others, and we navigate through life trying to make the most of what we have.
- *The growth mindset.* Our skills and talents are malleable. We each have unique strengths, but we also have the capacity to improve significantly in any area, over time, through effort and practice.

We will circle back to mindset a few times throughout the book. Here, I want to focus on how mindset affects our fear of failure. While the fixed and growth mindsets address our abilities, they are also directly connected to the fears of failure outlined by Conroy. For those who see our abilities as fixed, a failure is a judgment on some unchangeable aspect of ourselves. Since that trait is fixed, there is nothing that we can do about a failure except recognize our weakness—and probably feel ashamed or embarrassed. Ouch. In contrast, the growth mindset allows us to view a failure as a shortcoming, but only a temporary one (and sometimes just a shortcoming of luck). Our ability to grow, to increase our capabilities through effort and practice and then try again, provides a path forward after a failure, allowing us to mitigate the impact of the failure and reduce our fears.

While the concept of our mindset gives us a way to think about how to move on after a failure, what may be even more powerful is how it impacts the decisions we make before a failure ever occurs. Psychology researchers Andrew Elliot and Marcy Church have shown, for example, that fear of failure can lead to behaviors called defensive pessimism and self-handicapping.[5] Defensive pessimism involves setting very low expectations and predicting failure in advance. Self-handicapping goes even further, leading us to take actions that may actively undermine our chances of success. How often have you heard, thought, or said, "I don't even know why I'm applying. There's no way I will get that funding (or that job or award)." Or perhaps you've done something subtle in setting up an experiment that you know could sabotage the outcome. Why do we do these things? Because our fixed mindset means that a failure is a judgment on our abilities, and these behaviors offer us an "out"—an alternative way to justify or explain a possible failure. The problem is, they also make us more likely to actually fail, and we can all agree that's not what we're aiming for.

You may already have heard and read quite a bit about mindset and know that the common answer to this challenge is simply to "have a growth mindset." I'm not going to disagree, although I am going to push back on some of the ways this is messaged in society. Having a growth mindset does not mean saying, "I can do anything if I just try hard enough." Anyone who has heard me sing knows that no matter how much I might practice or work with a voice coach, it is unlikely I will ever be invited to perform in a musical. Instead, what a growth mindset encourages us to believe is that improvement is possible. Even if I never reach a professional level, my singing could still improve quite a bit if I decide it's something worth putting my time and effort into. Similarly, the more we work on building our writing skills or reading the literature, the more we will increase the likelihood of success with those grant applications. Another key aspect of the growth mindset is that beyond just "try harder" it can also mean "try differently." As we discussed in Chapter 2, there are significant benefits to leaning into your strengths. Thus, as you think about how to grow and improve in your grant writing, teaching, or other aspects of your professional life, you might need to try a different approach. This could be a relatively small adjustment, such as changing the process by which you read papers and draft a proposal. Or it could be a bigger pivot, such as realizing you are much stronger in collaborative environments and so you should focus on team proposals instead of individual ones.

## Dealing with the Pain

Another reason why failure hurts is because it takes away our sense of control. We want to pursue that research project, but if we fail to get the grant funded, we no longer control the choice to do so.

When a paper is rejected, we lose some control over where we want to publish. Moreover, as we cope with failure, it is not always clear when and how the painful feelings will subside. Not only can we feel a loss of control over our future, but we may also feel a loss of control over ourselves.

If you're wondering how I handled my first "not discussed," you can trust me when I say that it wasn't pretty. There were tears. There were swear words. There were swear words yelled through tears. There was an hour spent scrolling through job ads, because I was clearly not cut out for the faculty role and it was time to find a new career. I moped along through my day and into my evening. The one bright spot was a hug from my then three-year-old son, who had no idea what a federal grant was and couldn't care less whether or not mine would be funded. I woke up the next morning and begrudgingly got in my car to head back to work—to the job that I obviously wasn't good at. I arrived, started my day, and soon found myself talking with someone in my group about their research project, discussing data and what they wanted to try next. By the end of the conversation, something almost magical had happened—I felt normal again. Gone were the shame and despair. I was too busy thinking about this new project and the next grant application that we would be submitting. This was a huge "aha!" moment in my career, as I realized that: (1) it's okay to "feel the feelings," because the painful feelings won't last forever; (2) maybe I could even understand, anticipate, and replicate this process of moving on from a failure; (3) I didn't need to look for a new job after all. Since that day, I've weathered more failures than I can count. Sometimes I've bounced back quickly, and in other cases, such as my bad tenure vote, recovery has taken several months. What has been consistent is that understanding and owning my coping process has made a tremendous difference in both how I navigate failure and how I build the resilience to keep going after a failure, even when I know the next failure may be just around the corner.

Once you understand how to process your failure, you can retain control as you absorb and cope with bad news, and that can, in turn, make the coping process easier. It's also important to remember that every person is different and there is a wide range of coping strategies, so you need to experiment, observe, and reflect to figure out what is most effective for you and to understand the typical timeline for your process.[6] I initially leaned into a strategy referred to as "focus on and venting of emotions," which is noted by experts to be among the less useful strategies. In the end, the more positive strategies of seeking emotional support and planning for the future were what got me to a place where I could move forward. No matter what strategies work best for you, the important thing is to know your process. It's much easier to tolerate pain when you know when and how it will end. Moreover, I found that I could reduce the pain and accelerate the recovery process by heading out to the climbing gym for a bouldering session, spending time snuggling with my kids, or gathering my research group to brainstorm about future projects.

## On a More Positive Note

Okay, enough about failure. Let's talk about success! It's easy to believe that we don't even need to think about success—if things are working, then there's nothing to improve, right? Perhaps you are already an expert in dealing with success and don't need to think about how you manage it. I've personally found that although managing success is much more fun than managing failure, it requires just as much thought and intentionality. The three most common pitfalls I've seen and experienced when it comes to success are rushing past it too fast, dwelling on it too long, and questioning whether we've earned it.

## Slow Down and Enjoy

How often do you sit through a meeting that is mostly collegial and productive, but afterward all you can think about is that one harsh or sarcastic comment that was directed at you? Have you ever scrolled past twenty positive comments in your teaching evaluation but come away thinking they hated the class because of a critical comment from one student? Do you have a habit of spending hours thinking about a piece of bad news, and then when you receive some good news, you rush past it and back to your to-do list? If you think you're alone in doing this, know that you're not. This is a phenomenon called negativity bias, and while certain factors, such as depression, can make it more intense, it's an experience that is relatively universal.[7] The drivers of negativity bias are complex, but the reality is that failure demands our attention in a way that success does not. Perhaps we secretly fear that celebrating and enjoying our success will set us up to feel even more disappointed when our next failure arrives. Or maybe we're just so busy that we don't think we have time to stop and celebrate. No matter what the cause, we owe it to ourselves to pause and enjoy the moment when success comes our way. This could look like taking your group out for lunch when you get a grant funded, seeing a movie in the middle of a weekday afternoon to celebrate a manuscript submission, or reading over the positive comments in a review more times than the negative ones to counterbalance the natural cognitive distortion. Failure and bad news are inevitable parts of any job (and life in general) so we might as well enjoy the times when we're experiencing success.

Once we let go of our subconscious fear that celebrating our success will somehow taunt the universe into punishing us with a horrible failure, we can realize that celebrating is fun! This is also where we need to remember that it's good to enjoy good things, but moderation is key. It's not too different from the way that eating one slice of cake makes for a great dessert but eating the entire cake will leave you with a stomach full of regret. If we stay focused on success for too long, it can change from being something that motivates us to keep going into something that makes us complacent and stuck. At first glance, it may appear that this is the opposite of the problem of rushing through success too quickly, but in fact, I would argue that they are closely linked.

Let's talk more about cake. If you eat the first slice in big forkfuls while focused on the movie that you're watching, there's a good chance that you will feel unsatisfied when you finish it, and you'll want to reach for another slice. In contrast, if you take your time to enjoy every bite, you're much more likely to feel satisfied after finishing that first portion. (Note: for the purposes of scientific rigor, I ran this experiment on myself while writing this chapter, though the relationship between attention and satiety is also backed up by actual research).[8] Cake is amazing (obviously), but what does this have to do with success? When we try to rush past our success or keep ourselves from enjoying it as much as we should, we continue to feel unsatisfied and it can be difficult to move on to the next goal. But when we intentionally carve out time to enjoy our success, it can naturally motivate us to get back to doing the work that we love.

## Reality Distortion Filter

The other major hurdle that many of us encounter when it comes to success is the feeling that we don't deserve the good thing that is happening and everyone around us knows it. Welcome to impostor syndrome. While initially observed in high-achieving women, imposter syndrome can be experienced by anyone.[9] Not only does it keep us from being able to enjoy our own success, but it also makes us paranoid about how others will react to that success. At its root, this feeling is caused by our tendency to interpret reality in a way that casts doubt on our own skills or accomplishments. Valerie Young, who has spent her career studying impostor syndrome, says that the majority of people—of all genders—struggle with some form of this self-doubt. In her book *The Secret Thoughts of Successful Women*, Young emphasizes that while the uniting theme of impostor syndrome is that it distorts our view of our capabilities, we each have a unique distortion filter. Since there are many different ways that we can distort reality, there is no single piece of advice for fending off impostor syndrome. But we can overcome it by identifying and correcting the specific thought patterns that lead each of us to doubt the validity of our success.

When we get to the topic of conflict resolution in Chapter 10, we'll focus on separating facts from stories and understanding how facts lead each person to craft a story that describes their perception of a situation. Meanwhile, we can use the same framework here to mediate the internal conflict that is impostor syndrome. Simply put, facts and stories are the way that we make sense of the world. We subconsciously collect data from what we see, hear, and experience, and then we use it to arrive at conclusions. This can be helpful and even necessary. For example, if you are staying in a hotel and you wake up in the middle of the night to the blaring of an alarm, you sit up in bed and realize that you smell smoke, and then you peek into the hallway and see lights flashing, you're going to use those facts to craft the story that there is a fire in the hotel and you need to get out as quickly as possible. In this case, the story could save your life. Sometimes, however, our stories are incorrect. What if there is no fire, but instead there was a heavy-metal concert in the hotel ballroom and the smoke machine got out of control and set off the fire alarm? This mistaken interpretation of facts is what's happening when we experience impostor syndrome.

Let's look at a specific example related to work. When I started my position as a postdoc, I shifted to a new field of research. The facts I collected during my first week on the job included:

- I don't know how to use most of the equipment in this lab.
- I don't fully understand most of the research that people presented in group meeting.
- When people talk about other research groups working in our area, I've often never heard of them.

These facts led me to tell myself, "I don't belong here. They selected me by mistake, and I don't have what it takes to be successful." Over time, though, I gathered more facts, such as that many of my labmates also had to learn how to use the instruments when they arrived. This made me realize that I was telling myself a distorted story. I was comparing my first day to someone else's 354th or 782nd day. Once I recognized this, I could craft a new story and tell myself, "I have a lot to learn, but so do most people when they join this lab. If I work hard and seek out help, I can build my knowledge and expertise just like they did."

We can also distort reality by not taking into account all of the facts, positive as well as negative. While the statements from my first week were true, I could also have said:

- I know how to use many other instruments that aren't commonly used in this lab.
- I may not understand everything, but I bring a unique perspective to each meeting.

- I'm familiar with the work of other research groups that my current colleagues don't know about.

Using these facts, I could have crafted a different story: "I may not have the same specific expertise as the other people in this group, but I have other useful knowledge to contribute and the skills to be successful here."

Whether we need to reframe our story or we need to take a broader look at the facts, it is possible to work our way past impostor syndrome. It can be challenging to do this alone, however. The next time you find yourself mired in the impostor thought pattern, try reaching out to a friend or colleague who can help you to craft a more positive story so that you can enjoy your well-earned success.

## ACTION ITEMS

- *Own your failure.* Choose a recent failure, such as a grant or manuscript rejection (if you're like me, you will have many to choose from). Plan a time at a group meeting or other team activity when you can share about that failure. Talk about why you think things did not work out as you had hoped, what you can learn from the experience, and what your plans are for moving forward. If the failure caused you to make a change in your goals or approach, share about how you arrived at that decision and what you are planning to do differently in the future.
- *Master your coping strategy.* Think about how you responded to your most recent few failures or rejections. What was your initial reaction? How did your feelings evolve over time as you processed the failure? Think about the things you did that were

not effective in helping you feel better, and what you did that was effective. How long was it before you started to recover? Create a plan for what you can expect and how you will cope the next time you have a failure or rejection. Write out that plan so you can refer to it when you're caught in the storm of your emotions.

- *Own your success.* Reflect on the last time that you felt impostor syndrome. What were the facts that you were using and what story were you creating? Is your story true? Is there another story you can craft from your set of facts? If needed, reach out to a friend or colleague to ask for help and perspective. In parallel with crafting a story in which you can own your success, be sure to celebrate that success! Open your calendar and plan out time for something that will bring you joy, whether that's a celebratory meal with friends or family, or an hour of quiet time to crawl under a blanket and read a book. Whatever your celebration, savor every bite of your success.

# 6

## RECEIVING FEEDBACK

## We All Need Some Coaching to Be Our Best

IMAGINE AN ATHLETE WHO PLAYED BASKETBALL for several years and reached the highest levels in the sport, and then suddenly decided to switch to playing baseball. But we don't have to imagine this, because it actually happened. In October 1993 Michael Jordan shocked the sports world by announcing his retirement from professional basketball, and soon after, he signed with a minor league baseball team. In *The Last Dance*, the 2020 documentary series about Michael Jordan's NBA career with the Chicago Bulls, Jordan shares how he suddenly found himself realizing he had much to learn and a new skill set to master. Sound familiar? At some point in the transition from grad school or your postdoc position to your faculty job, you probably realized that the career game you were playing had changed dramatically overnight and you were going to need a whole new set of knowledge and skills in order to perform at your best.

In Jordan's case, less than two years after his retirement, he again captured the sports world's attention with the announcement that he would return to playing professional basketball with the Bulls—with whom he went on to win three more NBA championships. While Jordan could pivot back to his original path, it's likely that reaching your professional goals will require you to embrace and remain in your leadership role, even though it means you have to learn a new skill set. The important point here is what makes these transitions possible. Jordan highlighted this characteristic when he famously said, "My best skill was that I was *coachable*." We may have a lot to learn to become great leaders, but we can get there by being receptive and responsive to feedback and coaching.

There is no shortage of feedback in our world, and in academia we may feel that it is coming at us constantly in the form of grant reviews, manuscript reviews, teaching evaluations, and more. The challenge is that much of this feedback is not what we most need to hear, and it's often not coming from the people we most need to hear it from. First of all, it is often summative rather than formative—hearing why a particular manuscript is being rejected does not necessarily help you to craft a better research project or communicate your results with greater clarity in the future. Moreover, nearly all of the feedback we get is about our research, leaving us with little guidance for addressing the gap in our leadership skills. Finally, there is a challenge in who we receive the feedback from. Although Reviewer 2 may have interesting thoughts to share about the data in Figure 6c of your manuscript, when it comes to growing as leaders, the people we most need to receive feedback from are those whom we lead and those who lead or mentor us.

I'll be the first to admit it—receiving feedback is typically not very fun. Few if any of us would ever say, "I can't wait for you to tell me what I'm doing wrong and where I need to improve!" But that is what we all need to hear if we want to grow and get better, and the people we lead are counting on us to do exactly that. There is likely nothing I can tell you that will make this easy, but I will share about why it's difficult, how we can make it a little bit less painful, and how we can protect ourselves and manage our emotions when the feedback isn't constructive. And I promise that receiving feedback does tend to get easier the more often you practice it. Just as professional athletes rely on coaches and peers to help them master their technique, we can look to those around us for coaching when it comes to mastering our leadership skills.

## Ouch! That Hurts

Even when feedback is constructive, delivered with sensitivity, and coming from someone who cares about you, it can still hurt. Why is that? The answer lies in many of the principles that we've already talked about.

*Feedback threatens our identity.* I've already admitted that I'm not good at singing. What I haven't told you is that for much of my childhood, I *thought* that I was good at it. I sang loudly at birthday parties and during church Sunday school. I even tried out for the sixth-grade choir at my school. (I was not chosen.) It wasn't until college that my friends told me that I was really not good at singing, and that was painful. Why did it hurt so badly? It hurt because it felt like an attack on part of my identity. I was being told that something I believed to be true about me was in fact false. An identity threat like that can also be magnified by one's mindset. At that point in my life I had a relatively fixed mindset, so hearing that I was bad at something felt like a negative judgment about an unchangeable part of me. It's only now that I've shifted to a growth mindset that I can find the humor in this, because I recognize that my lack of singing ability doesn't have to define me as a person, and I know I can always work to improve if I decide that singing is important enough to warrant my time and effort. It also helps that, at this point in my life and career, my identity is not strongly tied to my ability to sing in tune. In contrast, much of the feedback we receive at work touches on topics that are closely integrated with our identity and it thus presents a much greater challenge. In a study of college students, psychologists Alex Forsythe and Sophie Johnson found that for those with a fixed mindset, receiving negative feedback resulted in self-defensive behavior and a failure to improve in the area of weakness. In contrast, students with a growth mindset viewed feedback as a positive "destabilizer" that was likely to challenge them in a way that would lead to improvement. As you might expect, the growth-mindset students engaged more with the feedback they received and were more successful as a result.[1]

*Feedback threatens our motivation.* In Chapter 4 we looked at how autonomy, mastery, and purpose can be significant factors driving intrinsic motivation, the healthy type of motivation that inspires us to perform at our best and have fun doing it. Negative feedback can take a swing at each of these factors, delivering a triple punch to our motivation. In the most straightforward case, negative feedback can erode our sense of mastery, as it often conveys the message that despite our effort, we are falling short of expectations. Depending on how it's delivered and the content, feedback can also deplete our sense of autonomy and purpose.[2] For example, feedback may send the message that we shouldn't pursue a project as we were planning to, or that we're not making as much of a positive impact with our mentoring as we'd thought. Individually or in combination, such losses to our perceived autonomy, mastery, and purpose can leave us feeling less motivated to do our work and less effective in the work that we do.

*Feedback threatens our relationships.* Receiving negative feedback can invoke feelings of failure, and our brains are wired to tell us that failure threatens relationships. Remember David Conroy's five factors that drive fear of failure, which we discussed in Chapter 5? Three of those factors can have a direct impact on our relationships with others, as they deal with shame and embarrassment, social influence, and upsetting the people we view as important.[3] Sometimes the feedback we hear tells us that we have literally failed at something that is important to those around us—the grant we need to support our lab members is not being funded, or the manuscript your lab members coauthored with you has been rejected. Other times the messaging is more subtle, but we can still feel like a failure if we did not achieve a desired outcome. Say, for example, that we hear from a colleague that our presentation at faculty meeting was not well received, or a student in our research lab points out that we did not handle a conflict between group members particularly well. In cases like these, we may feel the immediate sting of the failure itself, but the more painful experience can be the feedback it delivers about our shortcomings. This is especially true if that feedback is coming from someone we view as important in our lives. As Conroy's research shows, this is because we fear that falling short will change the way that person sees us and perhaps even threaten their respect for us or the value they place in the relationship.

*Feedback threatens our schedule.* It takes time to ask for feedback. It can take time to meet with someone to discuss their feedback. It usually takes time to emotionally process the feedback and decide how to respond. And feedback can compel us to make changes in how we do things, which in turn can require time to plan and implement. As discussed in Chapter 1, time may be the thing that you feel like you have the least of, and it may be the resource that determines the extent to which you accomplish your goals. Thus, embarking on the journey of inviting feedback can feel like a threat to both your schedule and your goals. It is *my* goal to convince you that it will be time well spent, and that it can help you become even more effective in achieving your goals, or at least create a better experience for everyone around you as you work toward those goals.

As you open yourself to feedback from people who care about you and your growth as a leader, you're likely to start realizing that what actually shapes their respect for you is not what they tell you but how you respond. In the moment when you are receiving feedback, that may be the first time that you are hearing that person's view of your capabilities. It's important to recognize that this information is new to you but not to them—they've likely been thinking about their feedback for some time and just haven't had a chance to share it with you. The fact that they're sharing it now doesn't make it more or less true, nor does it necessarily mean they've changed their opinion of you. However, if you respond thoughtfully and then make meaningful changes based on what you've heard, over time that can increase their level of respect for you as well as your level of influence.

## Going Full Circle

Now that we've unpacked the reasons why receiving feedback is challenging and painful, let's talk about something that makes it easier: for the most part, you get to choose the people who give you feedback on your leadership journey. Unlike the reviewers selected by publishers and funding agencies to assess your manuscripts or proposals, when it comes to asking for feedback on your leadership or mentorship, you can reach out directly to specific people you trust.

How do you decide who to ask? That is largely up to you, but the key is to think full circle. In the organizational leadership world, this is often referred to as "360 feedback," and the idea is to capture the perspectives of people who relate to you in different ways. Your department chair may view your strengths and areas for improvement differently from your research group members, and your faculty colleagues may have another set of opinions. This is because each group interacts with you differently and experiences your words and actions in a unique way.

Often, those you should reach out to for 360 feedback can be distilled down to three groups of people—supervisors, peers, and direct reports. However, I would argue that there are many other groups of people who have useful perspectives. In the academic context, I recommend thinking about getting feedback from:

- Mentors and former advisors
- Your department chair
- Leaders of committees you've served on
- Your professional coach
- Colleagues at your institution
- Collaborators and other peers outside your institution
- Individuals involved in service efforts that you've led
- Research lab members
- Students in your classes

Depending on where you are in your career, you may not yet have people in all of these categories. If you are a graduate student or postdoc preparing for your future faculty career, your feedback circle may include your research advisor, peers in your lab or graduate classes, and any students you teach or mentor. And depending on your goals, you may be able to list additional categories that are relevant to you (for example, I can guarantee that as I write this book, I am going to get *a lot* of feedback from my editors). As you think through what your 360 feedback list might look like, I do want to take a moment to highlight one person on that list that you might not have expected—the professional coach. We started this chapter talking about sports and the importance of coaching, and although *coach* may not be a term that we use frequently in academia, it should be.

You're likely reading this book because you need a set of skills in order to do your job, and these skills were not included in your academic training. I hope what you've read so far and the chapters that lie ahead will help to remedy that, but if you want to keep growing your leadership skills throughout your career, it can be very helpful to seek feedback from people beyond the students and faculty you typically interact with. Much like athletic coaches, professional coaches aren't necessarily able to do what you do, but they can help you get better at it. Executive coaching is widely used in the business world for exactly this purpose, and there are some coaches out there who are specifically focused on working with academics.

Professional coaching is something that you typically need to pay for, though some institutions offer free coaching through their human resources (HR) department or offer funds to cover this expense out of a professional-development budget. Or your employer may let you reimburse the cost out of your start-up or other nonsponsored funds. Over the past several years, I've worked with a coach who has helped me to manage a diverse set of challenges, ranging from what to say in a difficult conversation that week to determining where I want to be in my career in five or ten years. She also gives me feedback, and what makes that feedback so valuable is that it comes from someone who is a leadership expert and a person who genuinely cares about my success. That combination is tough to find elsewhere.

Still not sure who you should be asking for feedback? Another good way to figure this out is to go back to the goals you outlined in Chapter 3. We talked about trying to set SMART goals, where the *M* stands for measurable. It's likely that many of your goals are easy to measure, like expanding your research group to a certain size or publishing a specific number of papers. But some of your goals (possibly the important ones) are more challenging to measure. For example, you may have the goal of becoming a better mentor, creating a more inclusive lab culture, or improving your writing. This is where feedback can help. For any of your goals that don't involve a quantifiable output, think about which people in your life are best positioned to help you measure your success, and then seek regular feedback from those individuals.

## Gathering the Data

Whether you're eager to dive in and start asking for feedback or you're still working your way through the dread, a remaining challenge may be logistics. How exactly do you go about collecting feedback? In a few cases, the feedback finds you—think of your annual evaluation with your department chair or supervisor. However, these formal mechanisms are actually among the least useful, as they don't necessarily tell you how you're doing relative to *your* goals.

If you like the idea of having someone else structure the feedback for you but want a more useful (and likely more friendly) option than your mandatory annual evaluation, you can look for an opportunity to participate in a 360 exercise. Although these feedback circles are most commonly used in the business world, they can also be found in academia. I've personally participated in two different 360 reviews as part of the leadership courses I've taken—one through my professional society and one through my university. I know many others who have done the same. When you participate as the focus of a 360 review, you provide the organizer with the names and email addresses of people you know across many of the categories I listed in the last section. (You're encouraged to reach out to those individuals first, to ask if they would be willing to complete a survey about you.) The organizer will send them the survey link, compile their responses, provide you with a report, and typically hold a structured session to walk you through what the data mean and how to use the feedback to identify your strengths and areas for growth. The benefits here are that you don't need to manage the logistics and you receive support as you process the feedback. The downside is that you typically don't get to choose the categories or skills on which you are evaluated.

I've found the information from 360 exercises to be useful, but my biggest growth as a leader has come from the more direct feedback I receive on a regular basis. I find that soliciting this type of feedback can be as simple as going to lunch with a colleague and asking them to be candid about what went well and what could have been better in the committee that I led last year. Or I may ask a mentor for their thoughts about where they have seen me grow the most over the past few years and where they think I have the greatest opportunity for improvement. Above all, the most useful source for me, personally, has been the feedback exercise I organize with my research group one or two times per year. This is probably the most helpful because they experience the effects of my leadership more than anyone else. And more than anyone else, they are counting on me to keep improving and be my best.

Like many things I do as a leader, my group feedback exercise started as an experiment and has evolved significantly over the years. In the first iteration, I posted a link to a blank online document and asked, "How can I be better as a leader and mentor?" The document sat empty for a few days, and then all of a sudden it was full of comments. Some were constructive, some were not, and all of the comments were a little hard to take. I had thought that I was doing pretty well as a leader, and I suddenly had to confront the reality that I had quite a bit of work to do. I'll talk more later about how I processed this feedback (even the unconstructive bits). For now I want to stay focused on logistics.

The first big thing I learned from the exercise was that it was important for people to have the option to be anonymous, but it was not particularly helpful for everyone to see what others had written. Moving forward, I changed the format from an open group document to an online survey form, so that only I would see the feedback from each individual. Second, I learned that if I wanted more constructive feedback, I should ask more specific questions. Currently, my survey questions look something like this:

1. My goal is to be an outstanding leader and mentor for our research group. What am I doing well and what could I do better to achieve this goal?
2. Based on feedback from last year, my plan for growth as a leader and mentor was _____. Where have you seen improvement and what do I still need to work on?
3. A challenge that we navigated as a group last year was _____. What did I do well in that situation and what could I have done better?

As a result of this exercise, I have made several changes in my leadership over the years, and my group has also offered numerous suggestions for improving the feedback activity itself, which we have adopted. First, they decided that they wanted to get feedback too (this was seriously one of my proudest moments as an advisor.) We now have a feedback process in which everyone creates their own survey and we all complete them for each other. We've put some safeguards in place to allow people to ask narrower questions in order to limit the scope of feedback they might receive, and we've done quite a bit of professional-development work together on how to give constructive feedback (a topic we'll tackle in Chapter 9). Also, while the exercise was initially anonymous, over time people asked if they could start signing their comments. Now the overwhelming majority of the group does this in any given round—in fact, so many people disclose their identity that I have to warn everyone that those who don't sign their name may not be as anonymous as they expect. Finally, we vary the frequency of our feedback exercise between once and twice per year, based on what the group indicates would be most helpful to them.

The feedback system I've just described might be great for you and your group (in which case, feel free to use it). Or you may prefer a very different approach. One of my colleagues uses a system in which his group meets together without him, and they all discuss their feedback on his leadership and mentoring. A member of the group is chosen to be the reporter, and that person synthesizes and anonymizes the feedback and then shares it with my colleague. As you think about what might work best for your group, the important thing isn't the specific structure of the process but the idea that everyone involved should have a say in how it works. Our feedback exercise works for our group because we've all contributed to building it over the years, and that process has, in turn, built the trust it takes for it to go well and to be helpful rather than harmful.

## The Right Foundation

You might have read right past it, but there was a small word in that last sentence that has a big impact and forms the essential foundation for any productive feedback process—trust. We've talked about the challenges of receiving feedback, and there are also challenges associated with offering feedback. Leadership and management expert Amy Edmondson studies teams of individuals working in high-stakes environments, such as medical professionals performing heart surgery. When the stakes are high, you might assume that people would feel compelled to speak up and ask questions or offer feedback if they see something that doesn't seem right. Sadly, that is not the case for all teams. Edmondson finds that even when a patient's health is at stake, a key factor determining whether people will speak up is the degree of "psychological safety" they feel exists on their team.[4] The concept of psychological safety describes the extent to which people perceive that they're in an environment where it is safe to engage in interpersonally risky behavior—such as saying something critical or questioning a colleague—even though it could compromise a relationship.[5] You and your research group are likely not performing something as high risk as heart surgery, but the concept still applies.

In her book *The Fearless Organization*, Edmondson outlines how psychological safety is critical to success across a wide range of workplaces, including tech companies, schools, and government agencies. Fear is the primary threat to psychological safety, and it can be intensified by authority—the more authority a leader has relative to the other team members, the harder it is to create an environment in which people feel comfortable speaking up to offer feedback. It's important to recognize that academia has a particularly intense power structure.

You may work hard to create a culture where your lab members are valued as colleagues, but as the faculty member, and especially if you are tenured, you still have a tremendous amount of power and authority. That dynamic is going to pose a constant challenge to getting the candid feedback you need to grow as a leader and mentor.

What can we do about this? One way to lessen the effects of the power structure is to admit your own weaknesses.[6] As faculty, we have the opportunity to do this on a regular basis, for example by sharing stories of mistakes we made when setting up experiments as a grad student. We can also do this in the context of our leadership. When gearing up for our regular feedback exercise, I usually share feedback that I've received in the past, and I intentionally choose comments that are constructive but very strongly worded. Doing this shows my group that I know I have shortcomings as a leader and that I'm willing to receive critical feedback without retaliating. It also provides a baseline for their comments: "Whoa, someone said *that* to Jen and she was okay with it." I've found that rarely do the new comments I receive come even close to the level of intensity found in the ones that I shared. But given the significant power imbalance, I know that I need to create a lot of safe space to receive even a little bit of candor from the members of my group.

Edmondson also notes that unless team members know that their feedback is needed and desired, they are unlikely to risk sharing it.[7] Thus, the other key thing that we can do as leaders is to make it clear that feedback from our group members is important and impactful. Before asking for feedback, you can empower your group members by explaining why their feedback is so necessary. Using whatever choice of stories or analogies you want, outline how you have a leadership job that you weren't trained for, and why their feedback is critical to helping you learn and get better at it. Your response after receiving their feedback should also reinforce the fact that you value their time and input. Simply making changes in response to what they have shared is a good first step. But such changes will be even more powerful if you communicate directly with your group about why you are making each specific change—or why you are choosing *not* to do something that has been suggested. I've found it useful to hold a special group meeting in which I share quotes or bullet points from the feedback I've received (I keep it all anonymous even when people have signed their comments) and talk about my responses. This creates clarity when I decide to make changes, and it is even more important when I decide not to act on someone's feedback. You don't need to do everything that people suggest, but acknowledging their comment and explaining why you are choosing not to act on a particular suggestion are critical steps in building trust. A reasonable person is willing to be disagreed with, but nobody likes to feel ignored.

You can't build trust and psychological safety overnight. Nevertheless, each time that you invite feedback, receive it with gratitude, and show that it has real impact, you build up a foundation of trust, and it will keep growing stronger with time.

## Yes, but . . .

When you invite feedback, you will likely receive a wide range of recommendations, some of which may even contradict one other. You don't have to agree with or follow every piece of feedback, but how do you decide, and what do you do when feedback is delivered in a less-than-constructive manner?

Let's circle back to the first feedback exercise I held with my group. Some of their comments were spot-on. In fact, those were the ones that were the most painful to read, as I knew that those individuals had identified a particular area where I was falling short and needed to improve in order to be a good leader for my lab. On the other end of the spectrum, one of the comments I received was both inaccurate and unconstructive—it was based on hearsay, got many facts wrong, and was more focused on attacking my ethics than identifying a problem. This underscores two challenges that can arise when processing feedback: either the *content* seems off or the *intent* does. How do we handle these situations without demolishing trust?

In my case, I knew that I wanted to meet with my group to talk through their comments and highlight the things I would change in response. This discussion was relatively straightforward—though definitely not easy—with respect to the well-aimed and constructive feedback, but I had no idea how to handle the less helpful feedback. I didn't want to ignore it, but I also couldn't admit to something that had never happened—that would have been unfair to me, and it also would have cost me the respect of the people in my group who knew the truth behind the specific situation in question. In a stroke of luck, between the time that I posted the online document to collect feedback and the scheduled debriefing meeting with my group, one of my mentors happened to point me in the direction of a book titled *Thanks for the Feedback: The Science and Art of Receiving Feedback Well*. Perhaps most encouragingly, this book has a second subtitle: *Even When It Is Off Base, Unfair, Poorly Delivered, and, Frankly, You're Not In The Mood*. Needless to say, I immediately purchased the book and started reading it voraciously.

Authors Douglas Stone and Sheila Heen explain that when we get feedback, we're usually hoping for data—concrete examples of something we've done well or done poorly, and actionable suggestions for how we can improve. They point out that what we usually get instead is interpretations—data that have been filtered through someone else's life experience, perceptions, and value system, yielding a piece of feedback that can be heavy on judgment and light on specifics. This brings us back to the concept of facts versus stories that we discussed in Chapter 5. No matter what vocabulary we use, it's important to recognize that while stories are subjective interpretations of data or facts, they are still based on how someone has experienced a real event, and thus there is often something useful that we can learn from them.

So, how do we find the useful nugget of truth in a morass of poorly constructed or insensitively delivered feedback? The first step is to give yourself the time and space to let your emotional reaction to the feedback subside. Then return to it and start peeling back the layers of interpretation in search of the data. First, give yourself permission to discard anything that is factually inaccurate or based on rumor, or that attacks you as a person rather than critiquing your behavior or actions as a leader. Next, try to step into the perspective of the person who delivered the feedback. Consider how they would view the situation they are describing and how your words or actions might have made them feel. It may be helpful to ask someone you trust and who is also familiar with the situation to give you their perspective. Just like peeling back the outer layers of an onion, this process can be stinky or tear-inducing, but it's ultimately necessary if you want to get to something of use.

In my case, once I pulled away the layers of hearsay and hyperbole in the comment, I saw that behind the hurtful words was a valid concern about the way projects were assigned to people in my group—specifically, the lack of transparency about how such decisions were being made. Realizing that I could extract this point without having to keep all of the other stuff was hugely empowering and gave me a productive way to approach the topic with my group. I was able to talk to them about how I assigned projects and then formulate a new approach that would provide greater clarity and transparency. Perhaps most important, opening myself up to feedback (even when it was poorly delivered) and engaging with my group to discuss their concerns and suggestions made me realize that the people I lead don't expect me to be perfect, just open to listening and willing to grow.

## ACTION ITEMS

- *Draw your circle.* Think about the people you currently receive feedback from, either formally or informally. Look at the list of people in this chapter and ask yourself if there is a group that is not well represented among your current feedback circle. For each group of people, plan how you will seek their feedback in the next three months—think about who you will ask and how you will ask them. Put action items related to this plan on your calendar or to-do list.
- *Investigate your resources.* Talk with colleagues or start looking around your institution's HR website to identify the professional development resources that are already available to you. Does your institution have a way to defray the cost of professional coaching? Does it subscribe to any online leadership resources that may be helpful? Are there classes or programs offered through HR that would allow you to participate in a 360 feedback exercise? Depending on your institution's benefits package, you may also be eligible to enroll in undergraduate- or graduate-level classes in leadership or management without having to pay tuition. Courses in an executive MBA program may be especially useful, as they are structured for people who already have a full-time job.
- *Start exercising.* If you don't currently have a mechanism for gathering feedback from your research group members, set one

up now. If you're the only one who will receive feedback, then you can choose the format for the exercise. If you want to make it a group 360 exercise in which everyone gives feedback to each other, then hold a lab meeting to discuss how it will work and find a process that everyone can agree on. After you conduct the feedback exercise, hold another meeting to debrief the group and discuss changes that people would like to make for the next time around—because, yes, you even need to collect feedback about how you collect feedback.

# II

# *LEADING OTHERS*

# 7

# OWNERSHIP

## Because Leading Should Be a Team Sport

GOING ON A ROAD TRIP can be a fun and fulfilling experience, whether you are driving across the country or just a few hours from home. However, if you've ever embarked on a road trip with friends or family, you know that while it is often a great experience, it is rarely an efficient one. There may be extended debates over where to go—to the mountains for hiking or to the beach for lounging? There will likely be discussions about what to pack or when to depart, and the fiercest contest may be over the decision of where to stop for lunch on the way. It would be far less complicated if you were going by yourself, but then you would miss out on the chance to grow closer to people you care about and make memories that you can talk about for years to come.

This highlights an important transition as we move from discussing effective self-leadership to how to effectively lead others. In the case of self-leadership, our task is to set individual goals and then manage ourselves as we work toward them. In the case of leading others, the task broadens, as we are responsible for helping an entire team of people set goals and then empowering and encouraging each person to perform at their best and work together with others as they pursue those goals. In Chapter 3, we compared the process of setting and pursuing goals to driving to a desired destination, and leading a research lab is essentially one very long group road trip.

In the case of leading your lab, the specific decisions you will need to make as a group will largely revolve around project outlines, publication goals, lab policies—and perhaps where to go for lunch. Your group's ability to achieve your collective goals will depend on the extent to which each person is aware of the goals and is brought into the decisions.

So, how do we create familiarity and buy-in for things like goals and policies? The easiest way is to ensure that everyone is involved in the process of crafting and implementing them. The critical first step is taking ourselves out of the center. Academic culture is traditionally based on the "sage on the stage" model, in which the faculty member is the expert in the room. Our expertise is certainly of value in our role as a leader, but even more important is our capacity to learn how to empower others and engage in shared leadership. Thus, as we work through the rest of this chapter, I want to encourage you to set aside your own ideas and preferences, hand over control to the people in your research group, and see where that takes you. Much like on a road trip, there will be some unexpected events, but those could also be what makes the journey fun and memorable.

## Driver Safety

If someone handed you the keys to a car and said, "It's yours," would you get in and start driving? That probably depends on how much you trust that person. Perhaps it's one of your parents, offering you their old car as you head off to college. Or maybe it's a complete stranger who just robbed a bank and is trying to con you into driving the getaway car.

Ownership of projects, policies, and practices in the lab works in much the same way. If we want the members of our group to be willing to drive the process of setting and pursuing goals and articulating lab culture, then we need to build trust and create a safe space for them to do so. When we do, we may get an added bonus, as workplace trust has been linked to higher productivity and higher quality of work.[1] Thus, putting time and attention toward building trust can pay off in both the setting and achieving of goals.

What is trust? It is tough to define in a single sentence. And my definition of trust may be different from that of the people around me. While some definitions of trust focus only on a calculation of risk and value, more complex examples highlight the influence of emotional and relational factors, such as respect, acceptance, and appreciation.[2] In other words, when we choose to trust someone, we are making a rational decision based on our own calculation of the potential costs and benefits, but this calculation is inherently swayed by the relationship we have with that person.

From this definition, you may be thinking that trust sounds similar to the concept of psychological safety that we discussed in Chapter 6. You would be correct. Leadership and management professor Amy Edmondson notes that trust and psychological safety have much in common, but she also highlights some key differences. One of these is time scale—decisions made based on psychological safety tend to have short-term outcomes, whereas decisions made based on trust tend to have longer-term outcomes.[3]

Let's talk about this in practical terms. Imagine you're having a group meeting to discuss a research project that's not going particularly well. If you make an error in outlining what you think the next troubleshooting experiment should be, the level of psychological safety the group feels will determine the likelihood that someone will speak up and offer an alternative idea. Each member of your lab will be calculating whether the benefit of having a correctly designed experiment is worth the potential embarrassment if they happen to be incorrect, or the risk that you will respond poorly to having your ideas questioned. In contrast, when a member of your lab offers to take over leadership of that project despite all of the challenges, they are engaging in a calculation with much longer-term consequences. On one side, they see the potential for their own growth as a researcher and the publication that will be generated if they are successful. On the other, they are assessing their risk if the project is ultimately unsuccessful despite their best attempts—you might lose patience with them, for example, or not offer such a glowing recommendation when they apply for jobs in the future. That decision is about ownership, and it requires trust.

This leads us to the question of how we build trust. One of the most important factors is our actions—when people in our group are making a calculation of risk and value, our past actions serve as a critical source of data for them. In the previous example, as the group member considered whether to take ownership of the project, they probably thought about how you reacted the last time they had a failed experiment or they told you it was time to let go of a project. Our actions and responses matter, especially when we are on the receiving end of bad news. We can also build trust by building relationships. Clearly, spending time with the members of your research group is important to relationship building and leadership in general, but what you say during that time matters more than the time itself. How often do you express appreciation for hard work or say that you're glad someone is a member of your group? Do you show empathy when a lab member confides that they are dealing with a personal emergency? Even when your busy schedule is crowding out meeting or social time, a simple email offering praise or gratitude can have a significant impact. Encouragingly, researchers have shown that we can build mutual trust by entrusting someone with a responsibility and then empowering them to be successful. So, while fostering ownership requires trust, it can also be a powerful force for developing trust.[4]

What are the areas in which we can foster ownership in our research group? The potential list is broad and differs from group to group, but there's a good chance your list will include goals, policies, practices, lab culture, and projects.

## Choosing the Destination

In Chapter 3, we outlined a process for setting your own goals. But who is responsible for setting the goals for your lab? Moreover, who is determining the overall research direction? While you may have been raised in an academic culture that answered these questions with "the faculty member, of course," I want to encourage you to think about what it might look like to transfer some of the ownership to your research group. I will be the first to admit that I started doing this mostly out of necessity and serendipity, not as part of an intentional strategy to foster ownership. However, after running the leadership experiment and seeing the successful results, these practices have become central to the culture and operation of our group.

Let's start with the topic of the overall research direction. Here, handing over ownership likely came more easily to me than to most. Rewinding to my days as a grad student and then a postdoc, I knew then that I wanted to be a faculty member, but I was convinced it was impossible. My self-doubt prevented me from even considering applying for academic jobs until the last minute, shortly before applications were due. By the time I decided to give it a try, I was far behind most of my peers. I quickly prepared a set of research proposals, hoping that they would be good enough to get me a job but knowing that they were probably not good enough to get me tenure. As a result, I found myself sitting in my office on my first day as an assistant professor wondering, "What are we *actually* going to do?" My answer (out of necessity as much as leadership) was to engage everyone in the group in the idea-generating process. I played a significant guiding role, but from the very beginning, every member of my group was encouraged to share their project ideas, and we ended up pursuing many of those projects over the years. We have since formalized this process through our regular lab retreats, as our agenda typically includes time to brainstorm ideas for the funding proposals we plan to submit in the next six months and the projects we hope to initiate in the next couple of years.

Handing over ownership of our publication goals was equally unplanned, and in fact it happened as a result of one of the retreats. In the early days of the lab, I knew how many papers we needed to submit for publication each year for me to stay on track toward tenure, and that largely guided our goals. However, as the group grew larger and I grew as a leader, I realized that the lead author on each project had much better insight than I did on feasible timelines for manuscript submissions. Thus, when it came time to talk about publication goals during our retreat, instead of me giving a presentation on the numbers, I set up a few easel boards and handed the group a stack of sticky notes. Each group member was tasked with thinking about the papers they were working toward, writing each title on a sticky note, and then posting it on one of the easel boards according to when they anticipated submitting the manuscript. My role in the process was to look over which papers were being posted and where, and to ask questions or make suggestions when it seemed like a plan wasn't on target. After the flurry of activity, we all sat back and started counting the sticky notes. We had some ambitious goals!

When we returned from our retreat, we hung the easel pages in the lab, and every time a manuscript was submitted, we "moved a sticky note" to celebrate. In the first iteration, we found that our timelines were a bit overambitious. Nevertheless, motivation ran higher when each person was working toward a goal that they had led in creating, rather than one I had dictated to them. This also unintentionally became one of the most positive things we've done for our lab culture. Prior to this exercise, there had been a sense of individual accomplishment whenever a manuscript was submitted. But putting all of the sticky notes in one place made it clear that there was also a group goal—we would be succeeding as a group only if everyone was succeeding individually. Under the new system, every instance when a sticky note was moved became a whole group celebration, and the result was a culture in which cheering each other on is now a natural part of our daily routine.

This example brings up one more way our lab has benefited from shared leadership—in group operations such as the planning of our retreat. As I mentioned in Chapter 2, there aren't really rules for how to have a group retreat so the options are virtually endless. Some of the structure you choose may be dictated by your preferences and the resources available. For our lab's summer retreat, we typically rent a cabin and go away for three to four days. But we often hold a second retreat in the winter that lasts only one or two days, when we just gather in a conference room on campus. While the ambiance may not be as nice as a cabin, it also takes up far less time and money.

Beyond choosing the format, the way you spend your time during the retreat can be customized to meet the needs of your group. We held our first retreat three years after we started the lab, and we mostly focused on troubleshooting the challenges we were experiencing with our projects and mapping out the immediate next steps toward each manuscript. I took the lead in planning that retreat. A decade later, we have a retreat-planning committee within our lab, and that group takes ownership of logistics such as choosing the location and collating the shopping list for food. We work together to plan the program by discussing which goals are most important for our lab and how to use our retreat time to achieve them. This results in a schedule of sessions that might include professional development activities, working on grant applications, generating new research ideas, developing standard operating protocols or other shared resources, and updating our website. Finally, the group members take the lead in facilitating these sessions—everyone contributes to the content and direction of the discussion, while a facilitator guides the process and helps to ensure that we achieve the goals set for each session within the time allotted. Under the shared ownership model, we have a program that not only better serves our group's needs but also gets everyone more engaged in the activities because they have the opportunity to be a leader rather than a spectator.

## Setting the Rules of the Road

If you've driven or ridden in a car in different states or countries, you know that each place may have different rules and customs for navigating traffic. Similarly, each research group has different policies and norms that guide how they operate on a daily basis. These may include calendars used to sign up for equipment, specific instructions on who places orders for shared consumables, and a chart for how samples are organized in a freezer. Unless you are an extreme micromanager, you are probably already fostering significant ownership among those you lead when it comes to setting these "rules of the road." My goal is to help you think about ways that you can increase that sense of ownership and empowerment.

The most straightforward way to communicate lab policies and practices is through some form of a lab manual. Here is where I will admit that for the first seven years, my group actually did not have a written document outlining our policies. This was in part because our lab was small and we frequently experimented with new ways of managing our operations. Mostly, though, I resisted having a lab manual because I didn't want to create a thick stack of paper that everyone just tossed in the bottom drawer of their desk and ignored.

When we were moving between institutions in 2017, I realized that it was the perfect opportunity to get a fresh start and a new and better level of clarity around our policies. I also realized that as the member of the group who spends by far the least amount of time in the lab, mine was the least important opinion on what our specific policies should be. So, I turned it over to the group. We carved out a few hours during our lab retreat and we started by listing all of the areas where a policy would be helpful—ordering, shared glassware, freezer organization, and so on. Then we briefly discussed as an entire group what the actual policy should be. Finally, we divided into teams and drafted language for each topic, read through and commented on one another's text, and finished with a first draft of a policy manual. My goal since then has been to return to the manual at least once a year to make edits. Often, some institutional practices have changed and we need to change our policies in response. But more important, editing the lab manual as a group ensures that everyone (including me!) reads it at least once per year, so it doesn't get *too* dusty.

If lab policies are the equivalent of speed limits and stop signs, then lab culture is like the unofficial rules for how to merge in traffic and when to honk your horn. Similar to our lab's policies, our lab culture was something that we had discussed frequently but until recently had never articulated in writing. In this case, it wasn't because I was worried that it would be ignored. I just had no idea how to approach the task. Then one of my mentors suggested that I read Patrick Lencioni's *The Advantage*.[5] In his book, Lencioni walks through an entire process for building organizational health, but of particular use for the topic of lab culture is his question, How do we behave? What is especially helpful is that the answer doesn't come from simply asking that question. Rather, Lencioni offers more specific questions to consider as a group, and together they point toward your shared values. Lencioni framed his questions around a company setting, and I've found that with some modification they can work well for an academic research group. The questions we use are:

- What do we do in our group that is different from most research groups?
- Think about the people who are highly valued members of our team—what makes them so?

- Think about someone who is an outstanding researcher but would not be a good member of our team—what makes them so?

The last two questions are especially illuminating. The key is not to focus on specific names (and definitely don't write them down), but to brainstorm about the qualities and actions displayed by those individuals.

This exercise results in a list of actions and characteristics, and there are many different ways to synthesize that list into a statement of lab culture. It may make sense to craft a short paragraph or a bullet-point list of values or phrases. Our lab decided to make a word cloud, which now graces our lab website and the coffee mugs on everyone's desks. Having these values written out is helpful when we need to recruit new members to the group or offer feedback to each other, as it gives us a vocabulary to use in these discussions. And it works so effectively because it didn't come from me.

This is an apropos time to circle back to the topic of trust. How much of our cynicism in academia, and in life, is a result of seeing values written out on a website, or hearing them outlined in a speech, that are not lived out in reality? Sincerity and trust are important considerations for all of the activities in this chapter, but especially for lab values and culture. As leaders, it's our responsibility to live out the shared values that our group has articulated, holding people accountable when they don't measure up to them and apologizing when we are the one who misses the mark.

## Driving the Car

Once you've started to share ownership of the destination and the rules of the road, the last step is letting someone drive the car. As with lab policies, you probably already work to foster ownership when it comes to individual research projects—it may even be part of your lab culture. But we can all get better at it. I know I'm still improving.

This is an area where the most important action we can take is often *in*action. If you are just establishing your lab, there will be techniques and practices that you've carried out hundreds of times, but that are new to those in your lab. Yes, you could probably do things better, but that's not the point. It's often said that leadership is not about getting things done; it's about getting things done through other people. So, what do you do if you walk through the lab and see someone setting up an experiment in a way that might not work? We may think that our only options are to stop them and offer advice or ignore it and let them learn through failure. But there is a better option—asking questions. In this case, you might say something like, "I'm curious about why you decided to set up the experiment that way." Sometimes the person setting up the experiment will admit that they're a bit lost and that your advice and guidance would be helpful. Other times, the answer I receive is, "Because I already tried it the way you suggested, and it didn't work." Either way, asking a question rather than dictating directions is a reliable way to offer mentoring without taking away ownership.

What happens in the lab is an important part of fostering ownership, but just as important is what happens outside the lab. Think about all of the aspects of managing a research project that extend beyond the experiments themselves—deciding on an overall strategy for getting the experiments done, applying for a grant to fund the work, outlining and writing up the results for publication, presenting the findings at a conference. There are no set guidelines, and what works best will vary for every research group and every group member. The important thing is to have the conversation about these processes and keep finding ways to hand over the keys. As I'll talk about later in this book, our goal as group leaders is not only to mentor researchers now but also to grow the next generation of leaders for the future. Fostering ownership is a critical first step toward this, and it will make you a better leader as well.

## ACTION ITEMS

- *Determine your destination.* Whether you use our method of sticky notes on easel boards, a shared online document, or another format, take a few minutes at your next group meeting or retreat to set your group's publication goals. As the advisor, you can *advise* on what you think reasonable timelines and goals should be, but as the leaders of their projects, each lab member should *lead* in making the final decisions. Find a way to display your group goals, and celebrate every time a milestone is reached.
- *Capture the "rules of the road."* If you don't yet have a lab policy manual, set aside some time to draft one as a group. If you do have a policy manual, set aside time for everyone to read through it, note where changes are needed or would be helpful, and then discuss as a group and make any necessary edits. Whether you go with printed copies or share the policy manual electronically, make sure that everyone knows where to find it.
- *Check your response.* When someone in your lab has taken ownership of a policy, process, or project and something doesn't go well, your response will have a huge impact on their future decisions about how much responsibility they want to have. For the next week, after each meeting in which you receive some

form of bad news from a group member, write down how you did respond and how you *wish* you had responded. Think about your ideal responses and prepare to use them the next time you learn that someone has done good work but it hasn't led to the hoped-for results.

# 8

# ENVIRONMENT

## Creating a Place Where Everyone Can Thrive

"RECALL A TIME WHEN SOMEONE WAS KIND to you in a professional life." This task is part of an education research workshop that I'm participating in, led by Mica Estrada, a social and behavioral scientist at the University of California, San Francisco, who leads nationwide initiatives to increase diversity in STEM fields. Estrada gives us one minute to recall an experience and write about what happened and how it made us feel. I suggest you try this also—put one minute on a timer and think about an act of kindness that you've experienced in your school or work life.

Who showed you this kindness? As the minute runs out in our workshop, Estrada polls the audience to ask how many of us thought about an experience when the person who showed us the kindness was someone who had more power or authority than we did. The majority of hands go up. Estrada uses this to highlight the fact that, although we don't necessarily experience more acts of kindness from people in positions of power or authority, when we do, they have a greater impact on us.

Now for the tough part: as group leaders, our acts of kindness (or unkindness) will be similarly impactful and memorable for the individuals in our research group, and this carries with it a tremendous responsibility as well as opportunity. One of the most simultaneously inspiring and terrifying aspects of being a leader is that we have the power to shape a significant portion of the positive and negative experiences that our group members have at work. It is inspiring because, even when change is difficult in the context of the broader academic system, we still have the ability to create the culture and practices that we want within our own lab. It is also terrifying because once we realize what is possible, we have a moral imperative to work toward a culture and set of practices that will allow every individual to thrive. And, once again, *we weren't trained for this*.

In this chapter we will focus on the principles of diversity, equity, inclusion, and justice (DEIJ). There is a wealth of excellent books, articles, and other resources that discuss systemic biases and inequities and offer insights into how to foster DEIJ at the institutional or societal level.[1] I encourage you to seek out and continually engage with resources on these topics, as knowledge and best practices continue to advance and evolve over time. I also suggest that you give priority to resources generated by members of groups that have been marginalized by these biases and inequities. The resource lists referenced will offer a starting point, but the topics most relevant for you will depend on where you live, your discipline, and the type of institution where you work. Another great source may be your institution's office for DEI or faculty development. The experts there should be able to point you in the direction of useful books and other media.

As you dive into this topic, it's easy to feel overwhelmed by the magnitude of the challenge or lost as to how to make a difference. Changing the way things work in society and academia is a complex problem that requires a multifaceted approach. But one way we can have a powerful impact as group or team leaders is by changing the environment in our own lab. So, while DEIJ encompasses a vast spectrum of important conversations and requires systemic changes, I want to spend this chapter addressing one very specific question: What can we do as leaders to create an environment in our research group that provides everyone with the opportunity thrive?

## Who Is "Everyone"?

Diversity is a property of groups, not individuals, and there are many different forms of diversity.[2] If you have participated in conversations about workplace diversity, the most immediate dimensions that come to mind may be gender, race, ethnicity, sexuality, disability, and socioeconomic status. These are all important aspects to consider when supporting diversity, and we'll talk about why in a little bit. But before we get there, I want to consider a broader definition of diversity, which is based on the recognition that everyone has a unique set of identities, personality traits, life experiences, and perspectives. We foster diversity and create an inclusive environment when we respect and value these differences and see them as strengths rather than weaknesses.

In Chapter 2 we discussed a variety of personality assessments. Although they each have a different format and use their own language to describe personality traits, they share a common underlying principle—that there is a virtually infinite number of equally valuable approaches to viewing the world and performing work. And while it may sometimes be frustrating to work with someone who has a very different viewpoint or set of preferences from ours, research has shown that this can actually lead to better results, as working on a team of people who have different perspectives not only opens the door to more creative ideas but also pushes us to be more discerning when it comes to our own ideas.[3]

While it's important to recognize that there are many axes of diversity, this is also where "diversity" conversations often get derailed. Yes, we need to acknowledge that everyone has their own unique experiences and identities and embrace those differences. And if we lived in a perfectly equitable and just society, that would be sufficient. But we don't live in such a world. Instead, we live in countries and work in academic systems where biases such as racism, sexism, and ableism result in injustice, microaggressions, stereotype threat, and other challenges that make it more difficult for some people to succeed than others. While progress has been made in recent years to increase diversity in academia, the data still reveal widespread inequities in important areas—including grant funding, publication rates, and test scores—that underscore the systemic biases against individuals having identities that are historically marginalized in their field.[4] We also need to be attuned to intersectionality, a term coined by legal scholar Kimberlé Crenshaw that describes how a person's marginalized identities do not exist in isolation but rather can combine to amplify the challenges that individual faces.[5]

In parallel with acknowledging that diversity can boost team performance, we also need to consider how it affects each team member. Does every member of the team feel safe and respected while engaging in the work? Will everyone receive their share of the credit for the team's achievements? If our goal is only to achieve equality, where everyone is treated the same way, then we will never actually create an environment in which everyone has the same opportunity to be successful. Rather, we need to recognize and actively work to overcome and dismantle the systemic challenges that disproportionately impact some individuals, and in doing so, provide an equal opportunity for everyone to succeed and thrive. That is justice.

As an example of this in the context of the academic system in the United States, a student who identifies as Black is likely to encounter challenges that white students do not. Some accounts of these challenges are chronicled on social media under the hashtag #BlackInTheIvory and include being confused for a member of the custodial staff, being singled out by campus police, and experiencing a lack of access to mentors who share their identities.[6] Socioeconomic status and access to academic preparation can also create advantages or disadvantages for students. Education scholar Anthony Abraham Jack has shown that low-income college students can find it difficult to fit in and feel like they belong, or to learn the unwritten rules of how to succeed in college, and that this can be compounded when students have also not had access to an elite high school education.[7] Negative experiences and their intersectional effects have been described for students of multiple identities and are especially prevalent among those who identify as being from a minoritized racial or ethnic group, having a disability, as women or nonbinary individuals, international students, and first-generation or low-income students. The challenges can range from threats to physical safety to inaccessible workspaces to a lack of belonging. Together, they make success more difficult. Further, the attributes and knowledge students have acquired from being part of a rich culture different from the majority culture are often ignored—or worse, viewed as a weakness—instead of being recognized as assets for that student. As a group leader, you may not be able to address all of these inequities or fully reward each person's unique contributions, but you may be surprised by the impact you can have through the culture and policies that you create and live out in your research group.

## What Does It Mean to Thrive?

My initial draft of this chapter had a title that ended with the word "succeed," but I changed it because success is not a big enough goal. The word *success* implies visible outputs and metrics such as papers, fellowships, or awards. These are important but not sufficient. Success fails to account for the experience of each individual as they work toward and achieve those accolades. As I often tell my group, no amount of success is worth it if it destroys you as a person or ruins your life. Well-being matters.

One of the reasons we're so drawn to the idea of measuring success is because it comes with metrics that are easily measured. As discussed in Chapter 3, although you may not view the default metrics as the only things that indicate success, they provide a starting point. Thriving, on the other hand, is much more difficult to define, let alone quantify. There is no consensus definition for *thriving*, but it can include performance (those classic metrics of success) as well as growth and well-being.[8] In essence, thriving takes place when each individual has what they need to perform at their very best and the opportunity to experience fulfillment, self-efficacy, and a sense of belonging while doing so. In a research lab, this means that someone can thrive even when all of their experiments are failing—so long as they feel supported in that failure and can see themselves growing and learning as a result of it.

Another important distinction between success and thriving is that success can occur in a vacuum but thriving depends on the surrounding environment. Becky Wai-Ling Packard is an expert on the impact of mentoring, and in her book *Successful STEM Mentoring Initiatives for Underrepresented Students*, she makes a comparison between academic environments and ecological environments. Packard explains that if we were to notice that a particular type of bird was not thriving in our local ecosystem, it might be because of some characteristic of the bird itself, or it might be because the bird's environment has been polluted by nearby factories, resulting in poor water quality and a reduced food supply. In the latter case, no matter how much we study the bird or even provide supplemental food and water, it will only thrive when we remediate the toxic environment. In academia, where power can be inequitably distributed and the disparity in representation grows greater as people progress up the professional ladder, when an individual is not thriving, most often what needs to be addressed is the environment, not the individual.

## Cultivating the Environment

At this point, you may be feeling overwhelmed by the enormity of the task. Of course, if creating an environment in which everyone can thrive depends on your being able to fix all the systemic issues in society and academia, then real change may feel unlikely. However, I want to again empower you with the idea that as a group leader, you have more influence than you may think, and creating a healthy environment in your lab can have a tremendous impact on the researchers in your group.

Before we dive into how to do this, it is important to understand that while you are the person *responsible* for the environment in your lab, you cannot be the only one *creating* it. Rather, building an inclusive environment is a team effort, and engaging your group members in the process is critical. Every member of your group has different identities, and even when you share an identity, you may have very different life experiences around that identity. For example, in talking with other women in male-dominated STEM fields, I've found that some were encouraged to pursue their interests from a young age, whereas for others of us, choosing a career in a STEM field was an act of defiance against the cultural gender norms with which we were raised. In addition, the gender norms that we each experienced were often specific to the location and community in which we grew up. Thus, it is critical to first recognize that every person in your research group is a unique individual with their own life story, and then to listen and seek to understand their experience.

Involving the students and postdocs in your group is also empowering and trust building for them, both of which are key to a healthy environment. Biochemist and advocacy leader Tamra (Blue) Lahom highlights the importance of moving beyond offering students from minoritized groups "a seat at the table" to creating space for them to have a voice in the room.[9] Using the faculty hiring process as a case study, she describes how involving graduate students can increase representation as well as build trust between faculty and students. Specifically, Lahom piloted a program in which students met with each candidate and scored them based on a rubric that included metrics for the qualities of mentoring and commitment to DEI that the students hoped to see in their faculty. These data were reported to the search committee, allowing them to gather better information on each candidate and to elevate perspectives from members of minoritized groups without having to overburden the small number of faculty who identified with those groups. Lahom explains that one of the most important outcomes of the process was how it provided students with agency and impact while also helping them to understand the complexity of the hiring process, which resulted in increased trust between faculty and students.

So, what does this look like at the level of your research group? Creating and maintaining a healthy environment relies on both culture and policies. While these are not perfect definitions, you can think of culture as being all of the unwritten rules that govern your behavior and interactions with others, and policy as being all of the rules that are, well . . . written. Both culture and policy are important, and when it comes to cultivating a healthy environment, they are mutually reinforcing. As an example, let's consider a topic that we will discuss in depth in Chapter 9—giving feedback. In a healthy environment, feedback doesn't have to lack candor, but it should be delivered in a way that seeks to build up each individual and help them do their best, rather than tearing them down and making them feel discouraged. As a group leader, you can create policies that support this goal. You might make it a policy, for example, that when giving feedback during lab meetings, comments should be aimed at helping each individual improve and should focus on their work rather than on them as a person. That policy is important, but it's not sufficient. If someone truly wants to tear down another individual, they can probably find a passive-aggressive way to do so while staying within the bounds of the written policy. Thus, it's also critical to establish a culture in which this behavior would be considered unacceptable. In this example, when the written policy and unwritten rules of culture are aligned, the result can be open discussions about how to best give feedback and continual growth toward doing so in a collegial and formative way.

We started talking about both lab culture and policies in the last chapter, and if you've worked through the action items, you already have a solid foundation to build on. The goal here is to interrogate this culture and set of policies and continually make improvements to both, to ensure that everyone in your lab has the environment they need in order to thrive.

## Interrogating Culture and Policies

Thinking first about culture, ask yourself whether DEIJ is embedded in the ways that your group interacts with each other—you may use the words diversity, equity, inclusion, and justice directly, or these principles may be implicit in related terms, such as respect and kindness. A few guiding questions to consider as you interrogate your culture are:

- Do our conversations seek to build up and encourage or to tear down and discourage?
- To what extent are we open to learning from the perspectives and life stories of people who are different from us?
- Do the formats of our daily activities (such as lab meetings) create space to celebrate the unique aspects of each individual or do they pressure individuals to conform?

As you think about these questions, it's important to consider how the culture affects every member of your lab. And, as we've discussed, it's especially important to consider how each question might be answered by the members of your lab who identify with groups that are historically marginalized in your research field. The simple ability to see one's identities reflected in the people around us creates a sense of belonging that not all individuals are able to benefit from equally. Moreover, belonging—or the lack thereof—can have compounding effects.

Identity and justice expert Terrell Morton studied the experiences of Black women participating in STEM undergraduate research experiences at a historically Black university. From interviews with the students about their past and current experiences in STEM, he found that both the demographics and cultural norms of STEM create an environment in which the identities of Black women are hypervisible. The extent to which that hypervisibility had a positive or negative impact on the students depended on the level of support they experienced in the environment. When the students felt that they did not belong, they employed coping strategies that led to further negative social, psychological, and physical outcomes. In contrast, situations that had affirming cues—for example, the presence of Black women leaders, or culturally relevant research—led to increases in both psychological safety and student engagement in the research.[10]

As we consider how to create more affirming spaces, Mica Estrada's research at UCSF highlights how we can leverage intentional acts of kindness to counteract the effects of exclusionary experiences. She and her coauthors write, "Kindness cues that affirm social inclusion are the antidote to dignity violations."[11] Thus, while words of affirmation and other acts of kindness are important for everyone, they are especially important for those members of your lab who are more likely to have negative experiences in society or academia owing to their marginalized and unjustly devalued identities.

Depending on your own experience in academia, you may not have seen many examples of positive affirmation, and there are certainly many ways that you can tailor your approach to your own personality. What is important is intentionality. You can make a significant impact by simply taking a moment at the beginning of a meeting with a lab member to say, "I'm so glad you are a member of our group," or by sending an encouraging email or text message to a student on the day of their candidacy exam or other important presentation. As we'll see in the next chapter, you can even amplify the effect of your positive words by including specific details about what someone is doing well and the positive impact that is creating. While offering these affirmations doesn't take much time, it often doesn't happen unless you make it a habit. Think about times that you can regularly incorporate kindness into your daily routine—perhaps by sending a kind note with supportive feedback to each lab member after they give a presentation in group meeting. As department chair, I recently ordered a large stack of custom-printed notecards and initiated a habit of writing a note of gratitude to a member of our department every Friday. I leave the note in their mailbox before I go home for the weekend so that it's waiting for them on Monday morning.

Just as you can interrogate your culture, you can do the same with your policies. And while having a policy section specifically focused on DEIJ is a great start, it's important to take a broader view and recognize that nearly every policy (or lack thereof) impacts equity and inclusion in your group. For example, if you don't have a defined policy for how group members request or notify you of their vacation time, this can create an inequitable situation in which students in the majority culture feel very comfortable taking time off for the holidays that are widely celebrated at your institution, but an international student may feel far less comfortable asking for time off for a holiday that is celebrated at a different time of the year, or an Indigenous student may avoid asking for time to engage in an important cultural ceremony.

Although the ultimate goal is to look at every policy for your group through the lens of equity and inclusion, there are a few places where you can have an especially big impact, and focusing on these topics can be a great way to get started.

*Harassment, bullying, and discrimination.* We will talk more about how to manage these situations in Chapters 11 and 15, but in the context of lab policy it's important to state clearly that these behaviors are not tolerated in your lab and to provide a roadmap for what individuals can do if they experience or witness them. Reporting the behavior to you should always be an option for individuals who experience bullying, harassment, or discrimination in your lab, and other options can include having a direct conversation with the individual who carried out the inappropriate behavior, initiating a mediated conversation involving another lab member, or going directly to a university office that can provide support. You can also include information for your group members on institutional resources, such as a university ombudsperson or the university-level office where students and staff can report inappropriate behavior (if you're in the United States, this is often your Title IX office).

*Hidden curriculum.* There's a ton you need to know in order to be successful in academia that isn't written down or taught in classes. These topics can range from knowing whether you can (and should) contact faculty before applying to graduate school to how to format a group meeting presentation to how to network at conferences. Most of us figure these things out with time, but those who get a head start because someone shared this information with them are more likely to make a good first impression and thus have opportunities that others don't. These first impressions are particularly important because they form the lens through which all other information about a person will be filtered. One area where my own lab members identified hidden curriculum was in our lab's rotation system. Students who are considering joining the lab spend a few weeks rotating through our lab and others, and then they decide which lab they would like to join, and we decide who we are excited to recruit. In our first few years of operating in this system, I wanted to provide maximum flexibility to students, so I offered them the chance to spend their rotation time in whatever way they thought would be most useful. Unfortunately, that format fostered an inequitable situation in which some students had much more information than others on how to make a good impression during their rotation. What made this particularly problematic was that access to this information was largely based on privilege—for example, being a domestic student, having parents or other family members who had attended graduate school, or having extensive undergraduate research experience. Recognizing this challenge, our lab came together to design a new rotation system that has a structured schedule for each week and clear expectations. We also decided to match each rotation student with a senior lab member who could answer questions and help them understand how to be successful in the rotation. The result has been not only a more equitable process but also a more enjoyable and effective one for everyone.

*Equitable division of work.* Just as you may have service responsibilities in addition to your teaching and research, your lab members also spend time on activities beyond research. Many of these are activities that they find fulfilling, such as serving on a DEI committee or participating in outreach or mentorship activities. It is critical to recognize, however, that members of marginalized groups are disproportionately asked to take on these tasks, or they may feel more compelled to do so in order to help create a healthier climate in the department or beyond. You can't give them more hours in their day, but you can ensure that the amount of time they spend on these activities doesn't have to come entirely at the cost of their research. Think about the other jobs that exist in your lab—perhaps maintaining instruments or ordering supplies—and consider whether individuals who perform disproportionate amounts of service outside of the lab could be offered a lighter load in these other areas.

*Mental health and well-being.* While we often consider DEIJ and mental health to be separate topics, they are inextricably linked.[12] Our sense of belonging is directly related to well-being, and thus if you're creating policies that advance inclusion in your group, then you are creating policies that support the mental health of your group members.[13] The converse is true as well: when we create policies that support mental health, we create space for individuals with diverse needs to thrive. Two specific areas in which you can do this are your policies for flexible work hours and personal leave. Explicitly offering flexibility for lab members allows them to take time away during the day for appointments or errands, which can create space for a student who is facing a mental health challenge to meet with a therapist or other professional, helping them to effectively manage their condition while also succeeding in their research. Similarly, offering flexible personal leave time can allow a group member to continue their research while managing the health needs of a loved one. As a leader, it is up to you to decide what level of accommodation you are willing to provide (beyond the existing regulations in your institution). What should be universal is putting these policies in writing so that everyone has equal access to the information and is equally able to use the policies in order to thrive.

## ACTION ITEMS

- *Interrogate your culture.* Go back to the three questions I suggested earlier in this chapter, or craft your own questions about your lab's culture. Then spend some time thinking about how you would answer those questions and how the individuals in your lab, with their different identities and levels of privilege, might answer those questions. If possible, go beyond imagining and make this either an individual or group conversation. Identify areas where your lab culture can be improved to increase equity, inclusion, and justice, and then discuss with your group how you can make this happen. Track your progress by taking note of the climate in your lab meetings over time. A good metric to monitor is how many different members of your group are participating in discussions. If most of the comments and questions come from a small subset of your group members, then there is space for improvement. As you work on this with your group, you may become more attuned to the culture and participation level in other spaces, such as faculty meetings and scientific conferences.

- *Uncover the hidden curriculum.* Ask current group members what they know now, since applying to graduate school or joining the research lab, that they didn't know before, that has been important for their success. You will get a lot of information, and that's natural. Being a student or postdoc is a learning experience, after all. Still, you will likely be able to identify some critical pieces of knowledge that you can outline in your lab policy manual for everyone's benefit. Additionally, consider creating a policy for peer mentoring or other activities within the group that can further unearth the lab's hidden curriculum.
- *Create space for well-being.* What are your policies for flexible work hours and time off? Think about (or ask your group!) whether these are sufficient to support individuals with diverse personal or family needs. If they are not, then consider whether you can modify the policies to improve them. And no matter what, ensure that they are written down where everyone has access to the information. While you're at it, take a moment to make sure that the contact information for counseling services and other mental-health resources is posted in your lab or written out in your policy manual.

# 9

## GIVING FEEDBACK

## Candor Doesn't Have to Be Unkind

A NEW PROJECT IDEA IS COMING TO LIFE BEFORE MY EYES. It's the second day of my lab's annual retreat, and one of our sessions has been focused on developing new methods to address an important challenge in our research field. A senior graduate student is sketching out his team's idea on an easel board, and I'm struck by the boldness and innovation of the proposed approach. I also know that it's destined to fail, as there is a critical flaw in one part of the method.

I want to say something, but I'm afraid. I know that creativity is fragile, and I don't want to discourage this team of students as they are thinking big and taking risks with their ideas. In the end, I decide to point out my concern. Why? Because one of the ground rules for our retreat is, "If you notice a flaw or weakness in an idea, you owe it to your colleagues to speak up, even if it is uncomfortable to do so." Who made this rule? I did. Who struggles the most often to follow it? Also me.

I am a conflict-averse person. I would much rather watch a movie that I don't like or eat a meal that I don't care for than risk upsetting or offending the people closest to me. While this trade-off may be okay at home, withholding feedback at work is not an option if I want to be an effective leader. Moreover, giving feedback doesn't always have to lead to conflict, and not all conflict is negative (and for the times that it is, we'll talk about how to handle those in the next chapter).

As I raise my hand and share my concern about the project idea, something almost magical happens. Immediately after I finish outlining what could be a fatal flaw, someone else jumps in with an idea for how to overcome it. This leads to a new set of potential pitfalls, but those are quickly met with new alternative options. After fifteen minutes of rapid-fire feedback and redesigning, we arrive at a new idea that is even more creative (and more feasible) than the one the team had started with. Turning that idea from a sketch on the easel board into a functional chemical biology method would take many more failures and iterations and multiple years of effort, but it did eventually lead to a publication that our group is tremendously proud of. And we never would have gotten there without candor.

The lesson I learned that day is that if I really do care about the people with whom I work, then I owe it to them to be candid in my feedback. As we talked about in Chapter 6, feedback is coaching—it's not about making someone feel good or bad, it's about helping them get better at what they do. In the example of my lab retreat, it certainly mattered how my feedback was formulated and delivered. But we can learn how to do this well, and even when we fall short, imperfectly delivered feedback is usually far better than feedback that is never shared at all.

## Balancing Your Roles

As research group leaders, our day-to-day role in working with our team is typically that of a coach or mentor—our job is to help each person grow and perform at their best. But there are also times when you have to formally evaluate a member of your lab, perhaps for a candidacy exam, an annual performance review, or a thesis defense. On those days, your role may feel more like that of a referee, as you may have to make a tough judgment call.

The underlying source of these somewhat conflicting roles is the difference between formative and summative assessment. The theory of formative assessment was described in 1989 by education scholar Royce Sadler as feedback that helps students to understand the characteristics of high-quality work, assess their performance relative to those standards, and improve in order to reach their desired level of achievement.[1] Sadler developed this framework in the context of classroom teaching, but it can easily be applied to other areas, such as research performance. When reviewing recent progress with a lab member, for example, formative feedback can involve describing what high-quality data would look like, helping that individual to assess whether their results meet that standard, and offering advice for how they can continue to improve their experimental design or data collection and analysis. In contrast, summative assessment focuses on capturing an individual's achievement level at a specific time, for the purposes of reporting or making decisions about promotion to the next stage of their education or career. Hence, whether in the classroom or your research group, summative assessments will generally be high-stakes events where the feedback takes the form of a grade or a passing or failing mark. You probably encountered many such events during your own educational journey.

The dual role of being both a mentor offering formative feedback and an evaluator offering summative feedback is relatively common in leadership, but the line between these responsibilities is especially blurred in academia. We can't fully separate these roles, but we can learn to manage the relationship between them. Education researcher Alice Man Sze Lau cautions that we can easily fall into thinking that formative assessment is good and summative assessment is bad, and advises that we should instead view each as having value and being connected.[2] One way we can do this as research group leaders is by focusing our formative feedback on metrics and areas similar to those that will be assessed summatively in the future. For example, when a member of your group is presenting their work at lab meeting and has an upcoming milestone exam, you can focus your questions and feedback on helping them to achieve the level of knowledge and preparation that will be expected in the exam. Additionally, even though growth and improvement are not the primary purpose of a summative assessment, we can still build them in. This could look like offering feedback after a thesis defense that helps your group member improve their presentation before going on a job interview the following week.

## Your Mindset Matters

We talked about the importance of mindset in the context of receiving feedback, as a fixed mindset can lead you to view feedback as a permanent judgment about who you are, whereas a growth mindset enables you to view feedback as information to help you learn and improve. When you are giving feedback, you cannot control the mindset of the person who is receiving the feedback, but you can control your own mindset, and that can make a huge difference. In a widely popularized 1964 experiment, psychologist Robert Rosenthal partnered with school principal Lenore Jacobson to provide teachers with test results that predicted which students were poised for significant gains in their learning and abilities that year. When Rosenthal and Jacobson checked in on all of the students eight months later, they found that the students for whom they had predicted big gains did in fact show the greatest improvements in achievement.[3] Only there was no actual test, and those students had been chosen at random. They later named this phenomenon the Pygmalion effect because it demonstrates how our expectations of people are mirrored in our behavior toward them, such that people respond in ways that can make those expectations a reality. Your mindset can also affect how you approach assessment. You may have noticed parallels between a growth or fixed mindset and formative or summative assessment. While they may not be directly linked, researchers have found some evidence that teachers who hold a growth mindset are more likely to view assessment as an opportunity to offer individualized feedback that supports formative assessment goals such as self-evaluation.[4]

This highlights a key link between having great self-leadership skills and being a great leader to others—the mindset that you cultivate within yourself can directly impact the success of those around you. While your mindset is something that is continually shaped over time, you can create small habits that enable you to approach giving feedback in the most supportive and productive way possible. For example, if you need to have a meeting with a lab member who has fallen behind on their project, you could enter that meeting focused on the shortcomings in their recent data or a perceived lack of effort on their part, or you can choose to think back to a time when they were doing their best work and focus instead on their drive and capabilities. Either way, the goal of your feedback will be to help them get from where they currently are to where you know they are capable of being, but the growth-mindset approach is more likely to lead to a conversation that builds their confidence and helps them to make the necessary adjustments.

That last example also highlights one of the key leadership lessons I've learned over the years—almost nobody underperforms intentionally. If someone is working at a level that you know does not match their actual abilities, there is often something else going on. It could be something relatively minor, such as a string of disappointing results that have hurt their confidence and motivation. But sometimes it is something major, such as having a close family member who is dealing with a serious health issue. No matter what the situation, the best way to find out is to ask. Before launching into your feedback, take a moment to say, "I know that you care about doing great work and I've seen what you are capable of, but lately your performance hasn't been at that same level. Is there something going on that I should know about?" In my experience, the answers to this question can be wide ranging and can move the conversation in any number of directions. If a lab member is struggling with a serious life situation, then the conversation shifts to focus on how to support them. If they are dealing with a lack of motivation, I typically share some of the tips we discussed in Chapter 4 to help them rebuild their drive. Or, if they really don't have an answer, we are still poised to talk about where they should be with performance and how to make a plan for getting there. No matter where the conversation goes, taking a moment to ask the question can help you identify the root cause of the underperformance, and it sends the message that you care and that you believe that individual is capable of growth and improvement.

## Where Is the Bar?

We've just talked about underperformance, and this begs the question: Where is the bar? How do we know whether someone is performing below, at, or above expectations? Early in my career, I addressed this by convincing myself that I knew what "success" looked like, and that anyone who was not meeting that standard was not meeting expectations. This was not good leadership. Perhaps most important, I was failing to recognize that each member of my research group had different privileges or challenges that made it easier or more difficult for them to reach the standard of success. As we discussed in Chapter 8, these could include technical factors, such as access to research experience before entering graduate school or life situations such as the level of support provided by family and friends. Additionally, even if my standards had been equitable, I didn't have systems in place to communicate them clearly to the members of my research group.

There are many areas in which you can and should set expectations, and they encompass both *what* gets done and *how* it gets done. For example, you can set expectations for productivity, work hours, manuscript writing, lab safety, collegiality, and teamwork. Let's take a closer look at an expectation that is common to most research fields—publication. If you've involved your lab members in your group goal setting, as we talked about in Chapter 7, then you are already moving in a good direction because you've jointly determined reasonable target dates for manuscript submission for each project that they are working on. To create even more clarity around this expectation, you can work together to map out a timeline for when they should aim to complete each of the key experiments for the publication. You can also discuss the details of how the work should get done, such as whether you expect them to reach out independently to the director of a core facility that has equipment they will need, and whether you expect them to generate the first draft of the manuscript. The important thing is that you make this goal setting a collaborative activity. This allows you to better account for each individual's unique situation and skills and create goals that are attainable for them, and it helps ensure that the goals and expectations are communicated and understood by both of you.

## Choosing Both Candor and Kindness

Let's say that you have clearly communicated your expectations and now need to talk with someone who is clearly not meeting them. How do you deliver this feedback? Early in my career, I viewed feedback style as a one-dimensional axis—I could be kind or I could be candid, and the fact that I viewed these as opposite ends on the axis tells you everything about my definition of candor. Then, animated movies changed my view completely. Shortly before the lab retreat that I described at the start of this chapter, a colleague had recommended the book *Creativity, Inc.: Overcoming the Unseen Forces That Stand in the Way of True Inspiration*, by Ed Catmull and Amy Wallace.[5] Catmull is a co-founder of Pixar Animation Studios—the movie giant behind blockbuster films including *Toy Story* and *Inside Out*. Throughout their book, the authors describe candid feedback as being the fuel for creativity rather than an extinguisher of it. They detail how each movie starts out in very rough shape, and how it is only through a series of "brain trust" meetings, where everyone shares their thoughts openly about what is working or not working and what could be better, that the stories become refined into the hit movies that so many people love.

I instantly noticed the parallel between their process for making movies and my lab's process for designing research projects. I also recognized that our lab had been holding back on feedback because we didn't want to be unkind to one another. A key moment for me was reading this quote from Catmull and Wallace: "Candor isn't cruel. It does not destroy. On the contrary, any successful feedback system is built on empathy, on the idea that we are all in this together, that we understand your pain because we've experienced it ourselves."[6] Shortly after reading the book, I created the rule about candor for my group retreat, and the quote I just mentioned still graces the slide that I show every year as we head into our own "brain trust" meetings, where we propose and critique research ideas to sharpen and polish the projects that we will pursue in the coming year.

If candor and kindness don't have to exist at opposite ends of a spectrum, then how do they relate to each other? Business leader Kim Scott offers my favorite framework for this in her book *Radical Candor: Be a Kick-Ass Boss without Losing Your Humanity*. Scott writes that there are actually two different axes that can define feedback, challenging and caring. She then uses these to construct a model with four feedback styles—we can challenge directly, care personally, do both, or do neither. To briefly illustrate the four styles that Scott introduces, I've listed each and contextualized with an example we encounter frequently in real life—you're out to lunch with a group of friends and notice that one person has a piece of food stuck to their face. The way you respond can fall into one of Scott's four feedback categories:[7]

- *"Manipulative insincerity."* You are neither direct nor caring. Not only do you avoid saying something to your friend, but you also take the opportunity to make fun of them behind their back by ensuring that everyone else at the table notices too.
- *"Ruinous empathy."* You are caring but not direct. You feel bad for your friend but keep silent, hoping that the piece of food will fall

off on its own. You don't want to say anything because you're afraid it will embarrass them.
- *"Obnoxious aggression."* You are direct but not caring. You speak up but do so in a way that makes your friend the center of attention. They bear the short-term embarrassment, but at least the problem is fixed.
- *"Radical candor."* You are both direct and caring. You discreetly nudge your friend and signal that they have something on their face. If others notice, you play it off by noting that it's something that happens to all of us.

You might recognize all four of these styles in your colleagues at work, and you may also recognize that you've used each of them at various times when giving feedback. This framework helped me to see that when I thought I had to choose between being candid and being kind, I was really just oscillating between obnoxious aggression and ruinous empathy—not a fun path for anyone around me nor an effective approach to achieving our goals.

So how do we find our way to radical candor? Let's circle back to the example of meeting with a lab member who is underperforming in their research. You are already combining directness and caring when you start the conversation with, "I know that you care about doing great work and I've seen what you are capable of, but lately your performance hasn't been at that same level. Is there something going on that I should know about?" This question clearly states what you need them to hear (their performance is not at the level it should be), but it frames this question in a supportive way (they are capable of great work, and you care about what is happening in their personal life). You can also build in more caring both before and during the meeting. When someone is underperforming, it is natural for your default reaction to be frustration—their failure to meet project deadlines may be causing you stress about your upcoming grant deadline or holding back other members of your team. But you are more likely to get the outcome you want (and be the leader you want to be) if you can get into a mindset of wanting the best for that individual. Before the meeting starts, take a minute or two to remind yourself why you genuinely care about that person and their future. Sometimes this is easy, as they may have contributed significantly to your team in the past and shown great potential for the future. Other times, this can be more difficult. In those cases, your only answer to why you care might be because you're the leader and it's your job to help this person be their best. Once you're in the mindset of caring, start the meeting with the question we've queued up. As you proceed into the discussion, you can continue to build in caring while being direct by framing your feedback around how improving their performance will help them. You might say something like, "I know you have big goals for your future, and I think you can achieve them. However, some of the choices you are making right now are getting in the way of your doing that." You've communicated that their current performance is not meeting expectations, and you have done so in a way that sets them up for the outcome of getting better rather than giving up.

## Formulas for Feedback

The examples thus far illustrate how to be direct about a general problem, for example underperformance in research. Thinking back to Chapter 6, where we discussed the topic of receiving feedback, we know that the most helpful feedback is typically based on specific facts and events. If you're not sure how to deliver that type of feedback to someone else, you're in luck—there are formulas for this!

One of the most effective formulas I've found is the Situation-Behavior-Impact (SBI) model from the Center for Creative Leadership.[8] Part of what makes this approach so effective is that it's easy to remember and follow, especially in the midst of a challenging conversation. Delivering SBI feedback is as straightforward as the words sound.

- *Situation.* Outline when and where something happened.
- *Behavior.* Describe what the person did, making sure to stick to the facts rather than your interpretation of them.
- *Impact.* Tell the story of the impact their behavior had on you, or what you noticed about its impact on others.

In our example of the underperforming lab member, this could sound like, "When we met in my office last Thursday and discussed the grant progress report that was due the next day (situation), you said that you would email me the cell imaging data I needed by 5:00 p.m., but I never received it (behavior). I had to submit the progress report without those key data, which puts our future funding at risk (impact)." An additional step I often take to follow up on SBI feedback is to offer a suggestion for how things can be done differently in the future. In this case, you might stress how important it is to prioritize tasks that have important deadlines set by funding agencies. You can also say that you understand that unexpected challenges may arise, and in these situations, it is critical that they let you know as soon as they think that they might miss the deadline so you can work out an alternative plan together. After delivering SBI feedback, you can keep the conversation going by asking if they have a different interpretation of the event, or why they behaved in the way they did. No matter where the conversation goes, using this model helps you focus on the facts of the situation while also conveying your story of the impact that it had. And providing feedback in this way may also encourage the people in your group to adopt the same model when giving feedback to you!

Because we confront tremendous challenges with mental health in academia, you may fear that providing critical feedback to the members of your lab will exacerbate those struggles. In fact, the opposite is probably true. While it is not pleasant to hear that we are falling short or need to improve, researchers have found that uncertainty can be far worse. In a study of employees across two continents, psychologists Michael O'Driscoll and Terry Beehr found that uncertainty regarding job performance and doubt about whether the expectations of a supervisor were being met led to higher psychological strain and lower job satisfaction.[9] And academia is structured to amplify this effect. In comparison with the business world, which has clear metrics for success, such as profits and customer satisfaction, the metrics of success in academic research can be far more subjective and less quantifiable, making it more difficult for your group members to assess their performance. Thus, even if we don't enjoy giving feedback to the members of our research group, or they don't enjoy receiving it, providing direct, caring, and fact-based feedback is an important component of helping each person perform at their best and maintain their well-being.

You can also minimize stress and maximize the benefit of your feedback by customizing your approach to each lab member's preferences. As an example, if you are giving feedback on a thesis defense practice talk, you can ask whether the individual presenting would prefer to hear the feedback in real time during their practice run, hear it in an individual debriefing meeting afterward, or receive a compiled list of comments via email. If you're not sure what format they would prefer, take a minute to ask.

## Don't Forget the Praise!

As you employ these practices for giving feedback, remember that not all feedback needs to be about areas for improvement. Feedback can also be used to help someone understand what they are doing well. In fact, psychology studies focused on workplaces and educational settings have suggested that individuals perform best when they get at least a three-to-one ratio of positive to negative feedback.[10] And it's not just the amount of praise that matters but also how it is delivered. While it may feel good to receive a comment of "great work" or "nice job," and this type of praise may raise someone's motivation and sense of belonging in the short term, nonspecific praise misses out on the true power of feedback—helping someone to grow and get better at what they do. Just as with critical feedback, it's important for positive feedback to be specific. What exactly did someone do that was outstanding, and why did it make a difference? This is where the SBI model can again be very helpful. In the case of praise, an example of SBI feedback could be: "Last week, when our fluorescence plate reader stopped working (situation), I noticed that you took the lead on reaching out to the vendor to troubleshoot and coordinate repairs (behavior). I admire the initiative you showed, and your taking the time to do this helped our entire team by getting the instrument running again quickly (impact)." Not only is this type of feedback more fulfilling to receive than a simple "good job," it also helps that person understand exactly what they did well so they can repeat that type of behavior in the future.

When it comes to praise, there is also a significant benefit to offering it publicly. If you want more people to replicate the behavior that had a positive impact, consider sharing positive SBI feedback during a group meeting or other gathering of your lab. This magnifies the recognition for those who are doing good work, and shows your entire team the types of behavior that you value and applaud.

A few years ago, the members of my lab had an idea for how we could take this one step further—group awards. We now have two categories of awards: one is peer nominated and one is chosen by me, and both are given out two times each year. There are many different ways that you can structure group awards, but we have found that there are three things that make this practice positive and effective for our lab. First, we align the award criteria with the things that we value most. Rather than just giving awards to the people who publish the most papers, we base the awards on qualities such as perseverance, kindness, collegiality, drive, inclusivity, and courage. Second, the nomination statements focus on specific examples of when the individual demonstrated those qualities and the impact that it had on their work or the entire group. Third, I make the final decision in choosing the award winners. This allows me to focus on giving them out equitably over the years by ensuring both that we correct for biases and privilege, and that it is not the same small group of people who are recognized every time. At the end of each semester, I compile the nomination statements and read them at group meeting as I hand out the awards. The feedback I've received is that this practice is good for team building and helps people feel appreciated, both of which are positive outcomes. What is perhaps even more powerful is that it also reminds me to offer praise more often. It's easy to get caught up in the constant churn of deadlines and emails and forget to notice the great things that people are doing all around us. Yet we can be intentional about stopping to take notice and recognizing good work when we see it. Your positive feedback might just have an even bigger impact on performance than your critical feedback, and it will be much more fun to deliver.

## ACTION ITEMS

- *Check your expectations.* Before you dive into offering feedback on whether each member of your team is meeting expectations, take a moment to think about whether those expectations are equitable and whether your lab members know what they are. We can always benefit from more clarity, so take one step toward increasing your communication about expectations. This could be working with each lab member to create a timeline for their publication, or adding a section to your group policy manual about a topic such as work hours and time off.
- *Plan for candor.* Think about the conversation that you know you need to have but have been putting off because you fear it won't go well. You know which one. Write out your feedback in advance, using SBI format. You may have multiple examples, and each can have a separate SBI statement. Then think about why you care about that individual and what would be the best possible outcome for them. Finally, contact them and schedule the meeting.
- *Make someone's day.* Create a habit of noticing the things that people do well and offering praise. Think back over the last day or week. When did someone really stand out or go above and beyond? Using SBI format, write them a quick text or email, leave a note on their desk, or share your praise in person. Be sure to align your positive feedback with the behaviors that you value most and want to see more often. Repeat frequently.

# 10

## CONFLICT RESOLUTION

## What Do You *Really* Want?

MANY LEADERS EXPERIENCE A MOMENT when they realize that their job would be so much easier without all of the people. This is true. But then we would be unnecessary, as we would have nobody to lead. When each of us comes to work, we bring our personalities, perspectives, needs, and goals with us, and it is almost impossible for those not to clash with the personalities, perspectives, needs, and goals of the other people we work with. Thus, if working with a group of people is essential, then mediating conflict is going to be inevitable.

In the previous chapter, I mentioned that I am a conflict-averse person. This is why I completely panicked the first time I had to deal with a major conflict in my lab. While I did not handle that instance well, I later had the opportunity to learn some approaches to mediating conflict, and that turned out to be one of the most empowering skill sets I've ever gained. I wouldn't say that I *enjoy* dealing with conflict now, but I no longer feel the urge to hide under my desk when it arises, and I have even found that helping to resolve a disagreement can be one of the more fulfilling aspects of my job.

When we talk about conflict it's important to recognize that it can come in multiple forms. Even though I didn't use this term for it, the rapid-fire feedback on research projects that I described at the start of the last chapter can be considered "task conflict." As researchers, we are constantly pushing boundaries and exploring new territory, and we can do this much better if we are willing to disagree and debate ideas. The timing and context are critical, however—research has shown that task conflict is most effective in generating creativity when practiced in the early stages of a project, and that psychological safety is a key determinant of effectiveness.[1] When approached with both care and candor, conflict over ideas can prove to be a powerful process for refining a research proposal or improving a project plan. In contrast, interpersonal conflict can arise when the relationships between people become part of what is at stake in the discussion. The topic of whether interpersonal conflict can actually be healthy remains highly debated among researchers in psychology and organizational leadership, with some positing that when managed well, conflict can lead to greater connection and stronger teams, and others finding little beneficial effect.[2] The key is to not let task conflict turn into interpersonal conflict, and to work toward resolution when that happens.

If you can relate to my earlier story of dealing with conflict in my lab, and you, too, would rather hide under your desk than wade into a major disagreement in your group, then this chapter is for you. Conflict resolution is a skill that many professionals devote their entire careers to mastering. So, you won't become an expert overnight.

However, the following pages will help you get started and point you toward resources for continuing your learning and growth in this area. And while the examples in this chapter are focused on managing conflict in a research lab, you may find yourself using these skills at home or with friends, as well.

## Where Do You Stand?

Sometimes you can see a conflict brewing over days, weeks, or even months. Other times, you may receive an email or a meeting request that lets you know that a conflict has already occurred and needs to be resolved. And there will almost certainly be times when the conflict springs up in real time right in front of you. These situations offer varying amounts of time to prepare, but no matter what the details are, a good first step is to recognize the context for the conflict and where you stand in it. The most obvious dimensions to consider are your role and the relationships involved—are you one of the parties in conflict, and who are the other people affected?

Let's start with you. If you are part of the conflict, then mediating a dialogue can be significantly more difficult, as your emotions may be running high and the stakes will feel higher than when you are not directly involved. But it's not all bad news. If you are part of the conflict, then you also control a part of the process—what you choose to say and do. Resolving a conflict generally requires that all parties actually want to find a resolution, and if you are one of those people, then you are already partway there.

Now let's talk about the other people involved. As we've seen in previous chapters, it's important to be mindful of the power structure and the dynamics created by that structure. If you are a faculty member, you have a lot more power than a graduate student or research staff member. That creates an uneven playing field when it comes to managing conflict, especially if you are one of the people in the conflict. On the other hand, if the conflict involves colleagues who are at a professional level similar to yours, then this is far less of a concern. It's also possible that you find yourself in a conflict with someone who has power over you, such as your department chair, and that situation will be made more challenging by that power differential. When you are mediating a conflict between others, it's important to recognize the power dynamic that exists among all of the individuals involved. For example, if one person is a senior postdoctoral researcher who has known you for many years and the other is a relatively new graduate student who has been in the lab for only a few months, they do not have equal amounts of social capital in the situation.

As we discussed in Chapter 8, the goal in a situation of power imbalance is not to treat everyone equally but to achieve equity. If you are part of the conflict and in the position of relative power, this may mean that you will have to be the first to admit your shortcomings or negative contributions to the situation, as the stakes for doing so are much lower for you than for the other individual. If you are mediating a conflict between people with varying amounts of power, equity may look like first having a one-on-one meeting with each individual to hear their part of the story and to reassure the less powerful person that you understand the dynamics at work and you are committed to creating an environment in which they feel safe and supported.

Given how your position relative to others impacts the dynamics of the situation, before you even get into a conflict resolution process, it's important to ask yourself, "Am I the best person to mediate this disagreement?" If the situation involves your lab members and is arising from a disagreement related to their work, you will benefit from both knowing the individuals and having a strong understanding of the context, and this can outweigh the limitations imposed by the power differential. Of course, you may also be invested in the topic of the disagreement in a way that makes it difficult for you to remain neutral (or even to *appear* neutral, which is just as important). In these cases, it may be helpful to find someone who can either join you in the mediation process or take over and manage it entirely. As you weigh the decision about your role in the mediation, consider what resources are available at your institution. Do you have an ombudsperson? If so, then they can be a fantastic resource for advising you or managing the conflict themselves. Ombudspersons are generally appointed by the president or chancellor of a school to serve as a confidential advocate to whom individuals can bring their concerns or come for advice. They have training and significant experience in navigating conflict and likely will know what additional resources are available on your campus for managing the situation you are facing. When there is no ombudsperson or you prefer to talk with someone who is closer to the situation, you can also consider consulting with relevant leaders in your department. If your conflict is regarding faculty or staff, the department chair may be the best resource. For conflicts involving graduate students, you may want to reach out to the director of graduate studies or the person in the equivalent role for your program.

Importantly, involving others doesn't necessarily mean that you have no part in mediating the conflict. Much like a medical emergency, conflict can show up out of nowhere and may require both immediate and longer-term action. If you are in a public place and notice someone fall and break a bone, you might not be prepared to fully manage the situation, but you can recognize that you should call for help and try to keep the injured limb immobilized until the paramedics arrive. Even though you will not perform the surgery that ultimately results in healing the broken bone, you've still played a critical role. Similarly, in cases where conflict mediation ultimately requires help from someone like an ombudsperson to achieve resolution, you may be in a position to be a first responder. By knowing what to do and developing some basic conflict mediation skills, you can minimize the harm for everyone involved and play a valuable part in creating a restorative outcome.

## What Is Really Going On?

Whether by choice or not, you find yourself mediating a conflict. Where do you start? If you're like me, your instinct is to want to finish mediating the conflict as quickly as possible, so your natural response is to dive in and start problem solving in order to achieve that goal. One of the biggest mistakes you can make, though, is to wade into a conflict without having a thorough understanding of the situation. And that understanding often requires looking beyond the things that people are telling you.

The first question to investigate is whether there is an actual disagreement or if the individuals only *think* that they disagree because they haven't fully understood each other. Recognizing the difference between these two situations is critical, because resolving them requires vastly different approaches, and using the wrong approach can lead to more frustration than actual resolution. As a case study that we can consider throughout this chapter, let's imagine that two students in your research group are working together to write a manuscript, and a conflict has developed because one person is accusing the other of not doing their part of the work. It's entirely possible that the two individuals fundamentally disagree over what is a reasonable division of labor or timeline for completion—one person thinks that the other should be doing more of the writing than they are or should be doing it more quickly, while the other person thinks that the expectations being placed on them are unreasonable or unfair. And it's also possible that the two individuals have just miscommunicated. Perhaps one person said, "So, can you get me a draft of the manuscript by next Monday?" when what they meant was, "Can you get me a draft of your part of the manuscript by Monday?" Except the other person interpreted this as, "I expect you to write the entire manuscript and I'm telling you when it needs to be done." In this situation, it's easy to see how a conflict could escalate quickly even when the two individuals don't *actually* disagree on who is supposed to be writing which sections of the manuscript.

How do you figure out what is going on? Much like a detective, you need to ask lots of questions and consider the evidence as you try to piece together what actually happened to cause the conflict. You can also think of this process as similar to troubleshooting an experiment that isn't working—you gather data, formulate a hypothesis, and then find a way to test that hypothesis until you find an answer that aligns with what you are observing. In this particular situation, you might ask the student who is complaining about their colleague why they think that the other person is not doing their share. You could also ask whether they documented the division of work somewhere and how they reached their agreement about relative assignments. You would then also want to talk with the other student to understand their side of the story. Did they feel coerced into an unfair division of labor? Were they trying to get their part done but then a family emergency came up? Did they misinterpret an email from their colleague about deadlines, which caused them to get frustrated and give up? If you're not sure what to ask, you can also rely on general questions, such as "How did you interpret that?" or "What is your understanding of this?" and "How did that make you feel?"

As you gather information and piece together the actual situation, remember that every person is different. As we saw with the personality assessments in Chapter 2, we each have a different lens through which we view the world, and this affects everything that we say and hear. Think about how each of the individuals involved in the situation tends to behave, and how you have seen them interpret conversations in the past. As you listen to their answers to your questions, try to run them back through the filter of their personality to see what might be on the other side.

If you come to realize that a conflict is the result of a miscommunication rather than an actual disagreement, then the goal is to help each person see that and work toward a common understanding. This may not be as straightforward as it sounds, especially if emotions are still high or there is a history of conflict between the individuals. Still, you can get there, and a good place to start is by simply stating something like "From what I've heard, I think that you might actually agree on this. Perhaps we could talk it through in more detail." Alternatively, if you come to realize that there is an actual disagreement, then the path to conflict resolution will involve more steps, as you need to negotiate a mutually agreeable solution to the disagreement. But before we talk about those steps, there is a third possibility we need to discuss.

You may encounter times when your detective work doesn't seem to point to either of the situations I just described—there is no clear miscommunication and no disagreement, yet there is still conflict. In these cases, you need to think about what is *really* going on. Perhaps the conflict that is being outwardly expressed is just a proxy, and the real disagreement is about something much deeper below the surface. In the conflict over writing assignments and due dates for the manuscript draft, the disagreement could be a cover for a more sensitive topic, such as when one person feels that they should be higher on the author list but they are afraid to express that. It's also possible that the real conflict has nothing to do with the situation at hand—perhaps the two lab members are close friends, and one planned a fun activity over the weekend but did not invite the other. The seemingly unreasonable demand for a manuscript draft on Monday could be the slighted individual's attempt to get even while masking a personal issue as a professional one.

## What Do You Really Want?

If figuring out what is really going on is the logically demanding question, then figuring out what you (and they) really want is the emotionally demanding question. Just as you wouldn't hop on a city bus without first thinking about where you were hoping to go, you don't want to head into a difficult conversation without first articulating your desired outcome.

What makes this question so challenging is our innate motivation to protect our ego and return to emotionally comfortable territory. This can lead us to act as if what we really want is to be right, or to end the conflict as quickly as possible. Trying to navigate a conflict with those goals is unlikely to lead to the resolution that anyone is actually seeking. In their classic book on navigating conflict, *Crucial Conversations*, authors Joseph Grenny, Kerry Patterson, Ron McMillan, Al Switzler, and Emily Gregory recommend that before diving into a difficult conversation, we should stop and ask ourselves what we want—for ourselves, for the other person, and for the relationship.[3] They highlight that engaging in this process helps us get into the mindset of caring about everyone involved. Additionally, research has shown that cognitive tasks can aid in emotional regulation during conflict, so forcing our brain to work on a logic-based problem such as figuring out our desired outcome can help to bring our emotions under control, which in turn can help us to better navigate the difficult conversation.[4]

Going back to the conflict over manuscript writing, the desired outcome will depend on the underlying cause of the conflict. But both lab members presumably want to get the manuscript submitted as quickly as possible, want to write it in a way that maximizes the likelihood that it gets accepted, and want the author list to give fair credit to everyone who contributed—and hopefully they both want to continue to work together as colleagues. If you can unearth even one of these goals, you have identified an outcome that can generate a productive dialogue.

## Make It Mutual

When you have identified something that everyone in the conflict can agree that they want, you have taken the first step toward resolution. The authors of *Crucial Conversations* advise that as you work through difficult conversations, the two things you want to keep in view are mutual purpose and mutual respect. Let's take a look at mutual purpose first. This is exactly what it sounds like—a goal that you can agree on. In our manuscript-writing case study, it may be possible that everyone can agree that they want to write the best possible draft and submit it as soon as possible. Even if it seems that everyone agrees on this, it is important to pause and state it explicitly. When we're in conflict, our emotions go up and our ability to listen and understand can go down. Thus, you can make the ensuing dialogue more productive by first taking a moment to ask something like, "Can we all agree that we want to get a strong draft of this manuscript submitted as soon as possible so we minimize the chances that another lab beats us to publishing this research?"

If everyone agrees to the mutual purpose, then you're ready to start discussing the content of the conflict. Unfortunately, reaching agreement may not be that easy. As you get into the discussion of goals, you may find that one person is no longer invested in the paper and their highest priority is to complete a different project. In this case, you may need to widen your lens to find mutual purpose. As you talk through what each person really wants, look for commonalities—perhaps the two individuals don't care about the same projects but they still both care about maximizing publications for the lab as a whole, and that can serve as the common ground as you work toward a resolution.

If you feel like you're exhausting all of the reasonable options for mutual purpose and still not finding agreement, you may need to employ a different tactic. When emotions are running high and our pride or ego is on the line, we can be more hesitant to agree by saying yes because that feels like we are ceding control. In his book *Never Split the Difference*, former FBI kidnapping negotiator Chris Voss advises that it is much easier for people to agree to a no than a yes, so we just need to reframe the question so that we can reach agreement on what we *don't* want to happen. In our manuscript case, you might say something like, "If we don't work together on this manuscript, I'm concerned that someone else might beat us to publishing the same discovery. Do you want that to happen?" You've essentially asked the same question about mutual purpose, but now the ideal answer is the much more emotionally comfortable "no." If everyone agrees that this is not what you want, then you are poised to talk about what you do all want and to find agreement there.

Voss's approach also sets us up for my favorite question of last resort. When you have zoomed out as far as possible in search of mutual purpose and you're still struggling to find a shared goal, then a good question to ask is, "Do you want to be in conflict?" A reasonable and rational person would have a tough time saying yes, they do want to be in conflict. When your other options for mutual purpose have been exhausted, this question can help you find common ground in seeking a resolution, so that at least everyone can go their way and no longer be fighting. Asking this question can also help us recognize a critical challenge in conflict resolution: the small but real number of instances when one or more of the people involved is not actually willing or able to be reasonable and rational. We'll dig into how to handle this at the end of the chapter, but suffice it to say, if you ask, "Do you want to be in conflict?" and what you get is a blank stare, a change of topic, or a "yes," then achieving the outcome you want is going to require a different set of negotiation tools.

Okay, let's say you've identified the real source of the conflict and everyone has agreed to a common goal. You may feel like the work of conflict resolution is just getting started, but in reality the most challenging part is, thankfully, behind you. You're ready to dive into the logical work of problem-solving as you talk through potential options for achieving your mutual goals. If you are a researcher, then problem-solving is what you do for a living—your days may be filled with planning project strategies, deciding what to do when a paper or proposal is rejected, and troubleshooting failed experiments. This is why it can be so tempting to jump into conflict resolution by problem-solving right away rather than first building understanding. And once you do reach this part, it is also why you're likely to find that this is more comfortable territory than the previous steps.

As you discuss potential solutions together, the key is to maintain mutual respect. The authors of *Crucial Conversations* make the powerful observation that people usually don't get upset about the *content* of what is said. Rather, they get upset about what they perceive as the *intent* of the person who is saying it. As you negotiate with your two lab members to figure out how to get the manuscript written on time, one solution may be to redistribute the workload to provide some relief to the person who is falling behind. Keep in mind, though, that this suggestion could be met with a negative response if that individual believes they have lost your respect—that you are taking away their responsibilities because you no longer trust them or view them as capable. Successfully navigating this situation requires that you restore mutual respect, and you can do this using a skill called "contrasting." This involves recognizing what the other person may think is your intent, assuring them that is not the case, and then stating your actual intent in a way that highlights your respect for them. You might say, "I'm not doubting that you care about this project and that you would do a great job writing these sections of the manuscript. I do know that you're working on several important projects right now, and I want to help you balance your time so you can achieve all of the goals that are important to you." Maintaining mutual respect provides the guardrails to keep you on track as you keep discussing potential solutions until you eventually find something that is agreeable to everyone involved.

## Get It in Writing

Just as you would document the results of an experiment so your data and observations are not lost over time, once you've completed the hard work of conflict resolution, you want to put the details of the agreement in writing so you can refer back to them as needed. This can also ensure that everyone has a mutual understanding of what was agreed to. Even in a meeting where the topic of discussion is emotionally neutral, each participant can take away a different impression of what was said. This is amplified in a conflict situation, as research suggests that negative emotions can disrupt our working memory.[5] For this reason, it may even be helpful to take notes as you navigate a conflict, as your recollection after the fact will tend to be weakest for the most important parts of the discussion.

Whether you have reached a resolution or simply made progress toward one, take a little time after each meeting to write out key points from the discussion. You can use the framework I've described as a guide by outlining what you understand to be the source of the conflict, what was agreed on as the mutual purpose, and the practical actions that will be implemented to achieve this purpose. As you prepare to send this to everyone involved in the meeting, recognize that your memory is also imperfect. You can frame your report by noting, "This is my understanding of our conversation and the agreements we reached. Please let me know if I'm mistaken on any of these points and we can discuss." It can also be helpful to put a deadline on the feedback so that once that date has passed, the agreement is understood to be acceptable to everyone involved. The hope is that the conflict stays resolved and the written record fades into the background. If that does not prove to be the case and you need to revisit some points or deal with additional disagreements, then you have a solid foundation from which to start the next discussion.

## Congratulations—a Bonus!

As you learn and practice these conflict resolution skills, there is great news—you are gaining a bonus skill by also increasing your ability to negotiate. While we typically think of negotiation in the context of a job offer, equipment purchase, or other financial transaction, the conflict resolutions I have outlined are also negotiations. The main difference is the level of emotional investment involved. If you are trying to get the best price on a new instrument for your lab, that conversation is unlikely to be as emotionally charged as the authorship dispute that we've been discussing throughout this chapter. Another difference between a financial negotiation and a conflict resolution is the cost of walking away. A practice commonly taught in negotiation courses is to identify the BATNA—the best alternative to a negotiated agreement.[6] Exactly as the name implies, as a negotiation progresses, it is helpful to keep thinking about your best alternative option (and that of the person you're negotiating against) if either of you decides to walk away. In the case of the instrument purchase, you might have to go to a different vendor and perhaps pay a higher price or settle for a less ideal version of the instrument. In contrast, in an interpersonal conflict, the BATNA could be a permanently broken relationship.

Despite these differences, an important commonality exists between conflict resolution and financial negotiations: the core principles of mutual purpose and mutual respect. Whether you are trying to get the best price on a new instrument or asking your department chair for a raise, applying the skills and questions you've learned in this chapter will help you to be more effective in achieving your goals while also preserving—or even building—the relationship.

## Beyond Reasonable and Rational

Up to this point, we've been making the assumption that each individual involved in the situation is a reasonable and rational person who doesn't want to be in conflict with others. But there are exceptions. Some of the most frustrating moments in my personal and professional life have come from trying to navigate a conflict with someone I assumed was reasonable and rational, only to later realize that they were playing by a completely different set of rules.

Imagine being a player in a soccer game when someone else on the field is playing by the rules of volleyball. If you kick the ball to them and instead of kicking it back, they pick it up and spike it at you, your response will probably be a mix of confusion and frustration. And those emotions will escalate as the behavior continues. If you can identify what is happening, it won't necessarily make the behavior less annoying, but it will give you the ability to anticipate what is about to happen and to manage the impact.

How do you identify one of these situations? Sometimes all you have to do is ask. Several years ago, I found myself dealing with someone who had a habit of being hostile and hypercritical. As I tried to engage in a dialogue to resolve the conflict, they grew even more hostile. Trying to find mutual purpose, I said, "I don't want to be in conflict with you. Do you want to be in conflict?" Of everything I had said, this was the thing that left them speechless. The fact that they couldn't answer, "No, I don't want to be in conflict either," helped me see that no matter how good my dialogue skills were, it was unlikely that the conversation was going to result in the resolution I had been hoping for.

As I pondered what to do in this situation, and my instinct was to go on the offensive, I heard a piece of advice that brought me to a full stop. I was driving to work and listening to Sheila Heen, a law professor and member of the Harvard Negotiation Project, being interviewed on Adam Grant's *WorkLife* podcast.[7] Heen was talking about how to deal with difficult people at work, and she stressed that when confronted with someone who is behaving like an aggressive jerk, one of the worst things we can do is fight back. The part that struck me was her logic. Heen explained that if a person is behaving this way, they have probably been playing this game for a very long time and are an expert at being a jerk. If we respond by also being aggressive, we're choosing to play by their rules in a game in which they have much more experience than we do. Returning to our analogy, if you showed up to play soccer and that's what you're good at, don't let them draw you into a game of volleyball instead. Heen offered a practical strategy: helping that person to see the impact of their behavior on others and showing a bit of faith by saying that you think they are capable of finding creative ways to solve their problems without harming the people around them. This step can both help to resolve a conflict with someone who is overly aggressive and help you identify one of the most difficult circumstances in conflict resolution—when the person you are dealing with has traits of a personality disorder.

The prevalence of different personality disorders has been found to vary geographically and over time, although, interestingly, the overall frequency of diagnosable personality disorders in the United States has remained stable at around 10 percent of the population over a range of studies spanning multiple decades.[8] Data from the most recent nationwide US survey, the National Institutes of Health–sponsored National Epidemiologic Survey on Alcohol and Related Conditions, which included more than 40,000 participants, revealed that obsessive-compulsive personality disorder (OCPD), narcissistic personality disorder (NPD), and borderline personality disorder (BPD) were the most common disorders.[9] Whether someone has actually been diagnosed or just displays traits of a personality disorder, each of these sets of behaviors can present unique challenges to interpersonal relationships and conflict resolution. For example, OCPD typically involves traits of perfectionism or stubbornness, which can certainly make it more difficult to explore new options in search of a mutually agreeable solution to a problem.[10] Individuals displaying traits of BPD can be impulsive and emotionally volatile, which can derail a conflict-resolution discussion by effectively erasing any progress that has been made.[11] And while OCPD and BPD can make the individual steps of conflict resolution challenging, the traits associated with NPD can undermine the core principles of the entire process.

Debate is ongoing in the psychology community about how to characterize and diagnose NPD, as it can manifest in many different ways.[12] However, two nearly universal characteristics of narcissism are a lack of empathy and an altered view of reality.[13] In a situation where someone displays these traits, trying to use the normal approaches to conflict resolution can lead to more frustration than resolution. For example, if someone is not capable of empathy, they are unlikely to be able to engage in seeking a true mutual purpose that would benefit everyone involved. More frighteningly, narcissists often create an alternate reality to shield their fragile sense of self-worth, and they construct this reality from an intricate web of lies or exaggerations. When confronted with a fact that threatens their alternative view of reality, narcissists will react to protect themselves, and this can result in further aggression, deflection and reassignment of blame, and even more complex lies.

Sadly, there is relatively little you can do to work through a conflict with someone who shows traits of narcissistic behavior or has a narcissistic personality disorder. But recognizing the situation can help you set realistic expectations for the relationship and allow you to stop blaming yourself for not being able to work things out. Depending on the situation, you might also be able to respond with some empathy when you realize that the individual you're dealing with does not necessarily choose to behave that way (this doesn't make their behavior acceptable, but it may make it more understandable). Finally, you can create boundaries—a list of things that you will and will not allow to happen. Enforcing these boundaries can make someone with narcissistic tendencies very uncomfortable, and they may seek out other relationships that better support their alternate reality, offering you and the others involved a bit of respite. Finally, if the individual displaying narcissistic behavior is a member of your research group, then you have a critical responsibility to shield others in your group from their negative behavior and to create clear consequences that allow you to take action and ultimately remove them from your group if they continue to cause conflict or hurt others.

As a parting word of encouragement after a long chapter on a difficult topic, I want to emphasize that if you leave a conversation feeling like you haven't resolved the conflict, this does not mean that you have failed at conflict resolution. In addition to the caveats we just talked about, some conflicts (especially those with a long history) can take multiple conversations to work through, and sometimes difficult conversations need time to percolate before they produce a resolution. I've often been in a challenging conversation with someone that concludes with them raising their voice or walking away. In those moments, I used to feel like I had completely failed. In many instances, though, I found that the next time I saw that person, their attitude was completely different and the source of the conflict seemed to have disappeared. This made me realize that some people just need time to process. They are unlikely to move toward resolution in the moment, when their emotions are running high. But if you can have the conversation and then let them have time to think it through at their own pace, using their preferred strategies, they can develop an appreciation for the mutual purpose, restore their mutual respect, and then be ready to keep talking.

In the scenario I just described, sometimes the person raising their voice or walking away is me, and perhaps you can identify with this situation as well. In more than a few instances, taking a break from the conflict to go for a walk outside has been exactly what I needed to process my emotions and navigate back to mutual purpose and mutual respect. Difficult conversations are called difficult for a reason, and much like with a difficult research experiment, perfection is an unrealistic goal. Instead, we can develop our skills to perform at our best, learn from our mistakes, and cultivate empathy for the people with whom we are engaging in dialogue. That can be the recipe for ultimate success in restoring relationships.

## ACTION ITEMS

- *Assess the situation.* We spent this chapter going through how to resolve a conflict once it has become acute. As a leader, it is also wise to be looking out for conflicts that are just starting to form so you can address them before they get bigger. As you observe debates arising in your lab meetings or group lunches, and as you walk the hallways and see people interacting, take note of the interpersonal dynamics. If you sense an ongoing tension between two people or two groups of people, take the time to have an individual conversation about it. This doesn't have to be a formal meeting (and it may be better if it's not). Rather, find a time when you are already talking with one of the lab members involved, and ask, for example, "I noticed that you and [name] had some strong thoughts about each other's research during our last lab meeting. Is there anything going on that I should know about?" Sometimes what you saw is simply how those people choose to interact. Or it's possible that one person was just having a bad day and they had already talked about it after the meeting. But if there is a brewing conflict, they might share that information, and then you can start working on resolution. At the very least, you've indicated that you are open to a conversation on the topic.

- *Plan your conversation.* Think about a conflict that you are currently facing—it doesn't have to be a major argument, but maybe there is someone who you have had trouble getting along with. Think about what is really going on and what you actually want. What is your desired outcome for yourself, for them, and for the relationship? If you feel comfortable doing so, approach them for a conversation. If you don't feel ready for that step or you don't think that it will go well, talk about it with someone you trust, to get another perspective and advice for how to proceed.

- *Learn from experience.* No matter how skilled we become at conflict resolution and negotiation, there is always room for growth. After each difficult conversation, debrief yourself by thinking through these questions: What did I do well? What could I have done better? What did I learn from this? What would I do differently next time? Take a moment right now to think back to a recent difficult conversation and ask yourself these four questions, then make a habit of repeating this debriefing practice on a regular basis so that you can keep improving.

## 11

# ETHICAL LEADERSHIP

## Playing by the Rules

IF YOU WANTED TO LEARN HOW TO PLAY GOLF, you might start by reading up on the rules of the game, figuring out what equipment you need, and possibly even watching a few videos on how to choose the right club and the best way to swing at the ball. While all of this would give you important knowledge about how to play golf, it would not make you a golfer. Moreover, the type of golfer you become would be continually defined by how closely you follow the rules in each round that you play.

This example of learning to play golf highlights the difference between morals and ethics. While each of these terms has multiple definitions, one simple distillation is that morals are what you think is right and ethics are what you choose to do. In the case of playing golf, if you hit the ball into the trees and can't find it, the rules state that you are supposed to take a two-point penalty—knowing this is morals. But when you find yourself standing alone in the woods, you can decide to take the penalty, or you can just drop a new ball and pretend that it was the one you originally hit—the choice you make is ethics.

So, what does it mean to be an ethical leader? One of the most popular frameworks for thinking about this comes from management and organization scholars Linda Klebe Treviño, Laura Pincus Hartman, and Michael Brown, who highlight two components: the moral person and the moral manager.[1] Being a moral person involves adhering to principles such as making decisions in alignment with one's values and seeking to benefit others and minimize harm. Being a moral manager involves communicating those values to others and holding them accountable for ethical behavior. Excitingly, research shows that ethical leadership practices increase leadership effectiveness, and ethical leaders are able to create an ethical climate that in turn impacts the moral decisions of those who they lead.[2] This latter point is especially applicable to leading a research group, as it is impossible for you to be physically present when every experiment is run and each piece of data is collected. Additionally, there is a good chance that you will not have the capacity to review every analysis performed on those data. This is in fact healthy, as our job is not to do research for the people in our lab, but to mentor them toward becoming independent researchers themselves. Notably, this creates a situation in which the rigor of the research in each manuscript is directly reliant on the ethical choices of each lab member. Thus, while leading ethically is often something we do when nobody is watching, it has a broad impact on our effectiveness as group leaders and the reliability of our research.

Being an ethical leader may sound straightforward. After all, few people actively try to be *unethical* leaders. In reality, though, it is a challenging goal. This is in part because cognitive biases can prevent us from even being aware of our unethical behavior. Business administration scholars Max Bazerman and Ann Tenbrunsel discuss the impact of an unconscious bias called motivated blindness, which can lead us to do things such as skip over information that suggests we are making an unethical decision, if doing so will benefit us.[3] It is easy to see how this might apply to research—when we are reviewing the analysis of experimental results, there are many different ways in which data can be interpreted. Motivated blindness can cause us to subconsciously ignore samples or experiments that don't quite fit our desired outcome, or to see hypothesized correlations as being stronger than what the data actually support. In addition to the challenges introduced by our own biases, ethical leadership can be complicated by the wide range of values held by other individuals and cultures—my version of right may be wrong to you, and vice versa. Finally, benefit and harm are frequently linked, because a decision that helps one person may take away from someone else.

Although ethical leadership is complicated, it is worth it. In John Zenger and Joseph Folkman's studies of leadership effectiveness that we discussed in Chapter 2, they found that leaders can be successful using a wide range of competencies. Their work emphasizes that we each have unique strengths (as well as things that we are less strong at) and that this is okay. However, the one leadership trait that stands out in their model is "honesty and integrity." This is considered a core strength, and they note that its absence is one of those fatal flaws that can undermine the effectiveness of a leader no matter how many other profound strengths they have.[4] In essence, if you're only going to get one thing right, it should be ethical leadership.

## When the Greater Good Is Not Good

In our golfing example, the difference between good and bad seems fairly obvious. If you're playing golf and lose your ball, the right thing to do is to follow the rules and take the penalty. But what if there is more to the story? Imagine you are playing in a tournament and the other players plan to keep any prize money they win, but you have decided to donate all of your winnings to a charity that benefits children in your community. As you stand there in the woods, you are contemplating whether you should follow the rules and take the penalty, or skip the penalty because a better score will mean more prize money, which could have a tangible benefit for others. In this situation, one might say that following the rules overrides the potential to help others and so you should take the penalty. But what do we do in instances when our decision doesn't break any actual rules?

Many of the situations that we encounter in leadership involve finding a balance among multiple positive values, with no clear prevailing set of moral principles. As an example, let's consider the topic of assigning research projects and authorship on manuscripts. In the research community, it is widely accepted that if someone contributes data to a manuscript, they should be included on the list of authors. However, there are no universal norms for how many researchers should work on the project and thus be included as authors. Some labs are highly collaborative, with nearly everyone in the lab having a place on each paper, and some labs have individualized projects and publish papers with only the principal investigator and one or two other people as authors. So, how do you decide who to assign to each project? This decision can be guided by some practical aspects, such as how much time each person can spend and whether each person has a specific skill set needed for the project. Still, there is always an ethical decision to be made as you choose among the different options for dividing up the work. Additionally, as we get into these decisions, we may find that there are other important factors beyond time and expertise. Imagine that you are nearing publication and realize you need a few more experiments to be done, but they fall outside the expertise of the current authors. This is a great opportunity for someone to step in, complete a few experiments, and gain authorship on a paper. What if there are two people in your group who both have the right expertise, but the slightly less qualified individual is nearing graduation and doesn't have many papers, whereas the more qualified individual is already on a strong trajectory to graduation. Who do you choose for the project—the person who is most qualified or the person who will benefit more?

When a choice needs to be made and there is no clear moral right or wrong, many of us would instinctively appeal to the theory of utilitarianism, more commonly known as the greater good. This theory is popular for a reason, as it states that we should act to maximize happiness for the largest number of people. Sounds great, right?

I thought so too, at least until I took a bioethics class in college. In a very powerful lesson (evidenced by the fact that I still remember it more than twenty-five years later), our professor had us discuss the ethical frameworks we would use to make decisions. When many of us landed on the greater good, the professor offered us the following situation to consider. Imagine you work as a surgeon at a hospital and you have a patient who desperately needs a kidney transplant. A few rooms over, there is another patient who is there to have knee surgery. One patient will die without a new kidney, while the other has two kidneys but really needs only one. During the knee surgery, do you take out a kidney and give it to the transplant patient? After all, that would be for the greater good. This analogy bumps into some practical challenges with how hospitals and transplants work, but the point is clear—often the greater good is not what many of us would consider good. My professor explained that while utilitarianism is a valid ethical framework, it can impinge on another value—individual autonomy. Interestingly, a more subtle version of this ethical dilemma around organ donation is still debated in current medical policymaking, highlighting both the complexity of ethical decisions and their critical importance in real-world situations.[5]

Returning to the example of academic research projects, how do you ethically divide up work (and thus confer authorship) for manuscripts? The greater good framework would have you assigning each person in your group at least one experiment for each project, so that everyone is a coauthor on every paper, as this would maximize the good (publications) for the most people (lab members). However, it may be far from the most efficient way to get a project done, and it could also be seen as devaluing the contributions of the lead author, who will be doing most of the work.

While there are no clearly right or wrong answers here, this example highlights an important distinction between two different types of values—community-driven values and personal values. Community-driven values are the ones that are dictated to us by our government, employer, funding agency, or other entity that has some authority over us. In our publication example, the key community-driven value is the one dictated by the rules of research ethics (from funding agencies and publishers), which state that if someone collects data that are included in a paper, they should be included on the author list. In contrast, our personal values are those that we choose for ourselves. In the case of the publication, this is the value that determines how you assign group members to work on the different projects in your lab. Whereas community-driven values represent a set of rules that everyone in the community is expected to abide by, we each have our own personal values, for which there is no absolute right or wrong. As we will explore later in the chapter, ethical leadership relies not so much on what your personal values are but on your integrity in applying them equitably across different situations.

## Community-Driven Values

Depending on where you live, what type of institution you work for, and what field of research you are in, the community-driven values that apply to you will vary. Moreover, you probably received some mandatory training on upholding the values that are considered especially important—things like research ethics, lab safety, and dealing with sexual harassment. Those courses taught you morals, and how you put that knowledge into action as a group leader is ethics. I won't attempt to rehash all of the information that was in those training sessions, but I do want to elaborate on what it looks like to live out those values as a research group leader.

Research ethics is exactly what it sounds like—it affects all of the decisions you make in the course of doing research. Some of the rules that guide your decisions are universal across research fields—for example, you do not fabricate data or plagiarize someone else's work. Other rules will vary depending on whether your research involves mixing chemicals in a flask, testing a new cancer therapeutic in mice, or interviewing children about their experiences with failure. In the latter two examples involving people and animals, there is likely some form of internal review board that will look over and approve your experimental plan before you begin any work. I don't have space to cover the details of those processes, but do want to highlight the difference between being an ethical researcher and being an ethical leader. As a graduate student or postdoc, your responsibility was to be an ethical researcher—to learn the rules for your field and follow them. As a faculty member, you now have a responsibility to be an ethical leader, which means not only following the rules for your research field, but also communicating those rules to the people in your lab, providing visible examples of following the rules in your work, and holding your group members accountable to the rules as well.

An important aspect of ethical leadership in research that was likely never covered in a training session is the practice of rewarding well-aimed failures. No matter what your field, you're probably aware of at least a few high-profile cases of research misconduct in which papers were retracted or work was discredited as a result of falsified data. There are cases when an advisor was aware of the research misconduct (or was even responsible for it). There are also cases when the advisor did not participate in the misconduct, but it happened under their watch. The latter example should terrify all of us. As we discussed earlier, it is impossible for you to be present when every single piece of data is obtained and analyzed by your research group members, and thus your career and reputation depend to an inevitable extent on their ethical decision-making. It would be easy to say that there is never an excuse for unethical behavior. At the same time, for early-career researchers in a high-stakes environment, these ethical decisions can be difficult. Imagine being a fifth-year PhD student on a project that has never worked, and your advisor has said that you cannot graduate until you publish your work on the project. This is, sadly, a very real situation in too many labs, and it leaves students and postdocs with a difficult choice: whether to risk their career by falsifying data or to risk it by being honest about their failed experiments.

As an advisor, you can't make someone's research project work, but you can make it safe for them to be honest about the failures they encounter. In a study led by organizational psychologist and research-ethics expert Alison Antes, her team interviewed fifty-two group leaders who were nominated as "exemplars" for both their high-impact research and their integrity. Analysis of the interviews revealed several common practices that were credited with fostering high-quality and rigorous research in these labs. Many of them are topics we've already discussed, including sharing ownership in decision-making, ensuring transparent communication, and setting clear values and expectations. In addition, multiple exemplars highlighted the importance of creating a safe space for failure, with comments such as, "I tell [my team] that the only mistake that can't be fixed is the one they don't tell me about. . . . [I]f they've done something wrong, or they've discovered that something went wrong . . . I want to know about it. . . . I don't get angry . . . because everyone's human and we all make mistakes, including me," and "I let them know that progress isn't getting the results I want . . . progress is getting the results that the experiment gives us."[6]

As these quotes highlight, when someone is planning and carrying out experiments in a logical and sound way, that is what should define their performance, rather than the outcome of the experiments themselves. Moreover, we can demonstrate that we value these well-aimed failures by encouraging lab members to include them as part of the story they craft when they ultimately have enough results to publish a paper on their project. With projects that never make it to publication, the failed experiments and lessons learned can make for an interesting chapter in their thesis that shares this knowledge with the broader community. When we focus on rewarding the process instead of the outcome, and on supporting our lab members in failure rather than punishing them, we create an environment in which it is much easier for them to make ethical choices, and that benefits everyone.

Another challenging situation you are likely to encounter but were never really trained for is when a member of your group is experiencing bullying, harassment, or discrimination. Depending on where you live and work, the rules for dealing with these situations will differ, and you have probably received some form of required training on them. As an example, faculty in the United States are typically considered by their university to be mandatory reporters for Title IX violations. This means that if someone at work tells them about an experience of sexual harassment or sexual assault, the faculty member is required to report that incident to a designated person at their institution.

I will say more in Chapter 15 about how to support someone in your lab who is dealing with a situation anywhere on this spectrum of inappropriate behavior. For now, the key point that relates to ethics is recognizing your responsibility. When a situation falls outside of the official definition of Title IX or another legal clause, you have an ethical decision to make about whether to act. Perhaps one member of your lab is persistently spreading malicious rumors about another student in the program. Or maybe someone in your group made a disrespectful comment relating to another member's race, ethnicity, sexuality, or gender. In these moments, it can be tempting to tell yourself, "This isn't my responsibility" or even, "There's nothing I can do." On the contrary, we have quite a lot of power and authority (especially when it comes to our lab), and thus it *is* our responsibility and there is always *something* we can do.

As a first step, we can draw on the tools we discussed earlier for giving feedback and managing conflict, and we can have a discussion with the person who has been accused of inappropriate behavior. If that does not remedy the situation, consider the reward structure in your lab—what are the privileges that could be taken away if the behavior does not improve? This might include participation in group social events, the ability to keep working on a project or proposal, or even the ability to be a member of the research group. The important thing is to make it clear that the inappropriate behavior will not be tolerated, provide resources to help the individual correct their behavior, communicate the consequences that will result if the behavior doesn't change, and then follow through on the outcomes. And you don't have to go it alone. Antes and colleagues stress that seeking help is a key first step in professional decision-making.[7] If you're not sure how to handle a situation, you should reach out to someone such as a colleague, the director of your graduate program, your department chair, the human resources department, or an ombudsperson. No matter what the policies of our institutions and departments include, as faculty and group leaders we have more power than we may think. I would advocate that we have an ethical obligation to use that power to protect others from bullying, harassment, and discrimination.

The final category of community-driven values that I want to touch on is safety. Similar to research ethics, the specific safety rules that apply to you and your lab will vary depending on your research area. Also, the extent to which you follow these rules and create accountability to ensure that others follow them will have a profound impact on the safety practices of your lab. We talked above about how terrifying it is to imagine someone in your group participating in research misconduct without your knowledge. One of the few things even more terrifying is the thought of someone being permanently injured due to a safety mistake or lab accident. Thus, ethical leadership includes promoting safety rules, providing the resources needed for the people in your lab to follow those rules, and being clear about the consequences of violating the rules.

If you work in a research area that doesn't have safety hazards, you might think that this is a topic that doesn't apply to you. That's not the case. In addition to being responsible for the physical safety of our lab members, we also have an ethical responsibility to look out for their mental-health safety. Although we can't force someone to take care of their mental and emotional health, we can model healthy behavior and create an environment that supports the well-being of each individual. An entire book can be (and has been) written about mental health for graduate students—*Managing Your Mental Health during Your PhD*, by chemist and mental health advocate Zoë Ayres. Thus, I won't attempt to get into the details here, but I do encourage you to read up on the topic. The important thing for our discussion of ethics is recognizing that even if you are not a certified mental health professional, there is still much that you can do as a leader to protect the people in your research group.

- *Talk about it.* Unlike lab safety, there is still a stigma tied to talking about mental health in the workplace. You don't have to plan an elaborate presentation, but you can talk during your lab meetings about the importance of self-care and include a section in your lab manual that points people toward mental health resources and encourages them to take the time they need for their well-being.
- *Ask about it.* There are many times when it is impossible to tell that someone is struggling. But there are other times when you *know* someone is not okay. Take the time to ask how they are doing. If you're not sure what to say, you can lean on the tools for giving feedback and say something like, "I noticed that you haven't been participating in group meetings as much as usual and I'm wondering if everything is okay. I just wanted to let you know that I care about your well-being. Is there anything that you want to share with me or anything I can do to support you?"

- *Promote it.* The challenging thing about promoting mental health is that it's not as simple as providing access to counseling services or yoga classes. Every aspect of your lab—including group culture, professional development, and your leadership—has an impact on the mental health of your group members. This also means that if you've been working through the action items in this book to create a positive and inclusive culture, provide supportive feedback, and help people set reasonable goals, then you are already taking actions to support the mental health of your lab members. Keep up the great work!

## Personal Values

Unlike community-driven values, there is no absolute right and wrong when it comes to your personal values, because they are, well . . . personal. That doesn't mean, however, that they are exempt from being part of ethical leadership. We talked in Chapter 3 about your own values and goals, and then in Chapter 7 we discussed the values that you want to have as a lab. So, you already have a strong foundation when it comes to these personal rules. Where ethical leadership gets involved is in the ways you live out your values in your everyday conversations and decisions.

A phrase you may hear from leaders at your institution or elsewhere is "We need to align our actions with our values." I used to be a big fan of this statement, but more recently I've changed my mind. What I dislike about this framing is that it highlights the wrong problem. Our actions are in fact nearly always aligned with our values—they just may not be aligned with the values that we claim to hold. The real question, and the one at the heart of ethical leadership, is: Are the values we are displaying in our words and actions the same as the values that we say we have and that we want to have? As you might guess, there will be an action item on this.

## Consistency Counts

In addition to making sure that our actions are consistent with our values, we also need to be mindful of how our actions and values are consistent across different situations. We talked briefly at the start of this chapter about the importance of integrity, and you have likely seen a reflection of that characteristic in each of the ethical leadership situations we've worked through. Similar to the term *ethics*, there are many definitions for *integrity*, and I hope to convince you that it means more than just being honest or doing the "right thing." When thinking about integrity, it's helpful to turn to the definition that is most tangible—having integrity means being complete and undivided.

You may not consciously think about this, but every time you walk into a building, you assume that it has structural integrity—that all of the framing and joists and other supports are connected to form a stable structure and thus the building is going to remain standing. If you were to approach a building and see a gaping crack running up the wall or notice that the floor was missing, you would think twice before entering, as you would question its integrity. This building's potential for catastrophic failure is similar to what happens when a leader is divided in their actions around a community-driven value—if, for example, they are calling on their lab members to be rigorous in interpreting data but fabricating results on their own grant applications and progress reports. Much like the building that lacks structural integrity, entire careers can crumble quickly when someone lacks integrity in how they apply the values of research ethics, behavior among the group, and safety.

Lack of integrity can also be more subtle. When you ride a bicycle, you are counting on all of the parts to work together as a whole. If you were to remove a relatively small part, such as the chain or the pedals, it would still look like a bicycle but it would no longer have the integrity needed to be functional. This is what happens when we're not consistent with our personal values—for example, if we praise one member of our lab for their teamwork but criticize another person on that team because they're not working independently enough. Because such comments reflect personal values, there is no absolute right or wrong in each situation. Nevertheless, we still have an ethical obligation to apply our values consistently across different situations. Lacking integrity in your personal values may be less likely to lead to catastrophic career failure, as these values tend to have lower stakes than things like research misconduct. However, much like trying to ride a bicycle without a chain, this lack of personal integrity will render us ineffective as leaders, as the people we lead will see the inconsistency and quickly lose respect for us and what we have to say.

An important distinction to make when talking about integrity is that acting consistently according to your values means pursuing equity, not equality. In Chapter 8, we discussed how treating everyone equally doesn't necessarily lead to equity, because each person has a different situation and thus different needs. As an example of promoting equity, if your personal value is to provide each individual with the mentoring they need to be successful, you can apply that value consistently across all members of your group while varying the specific mentoring actions for each individual. Similarly, Max Bazerman points out that when the best interests of two people are at odds, we can maximize value while still acting ethically by focusing on what is most important to each person.[8] An example of this in academia might be when a student in the lab graduates but significant work is still needed to address the revision requests for their final manuscript in order to bring it to publication. Imagine there are two other students who have participated in the project and are still in the lab, and performing the experiments needed for revision would justify moving one of them to a co–first author position. Deciding which lab member should get this opportunity gives you a chance to explore what is most valuable to each individual. Perhaps one student is planning on an academic career, and thus the co–first author role on the paper would be most important for them, while the other is aiming for a career in science policy and is willing to remain at a lower contribution level now if they could have the opportunity to coauthor the next invited perspective article with you. In this case, by giving the immediate opportunity to the first student, you're not treating everyone equally but you are being equitable in seeking to maximize the career benefit for both, and thus you are acting consistently according to your personal value of helping each individual pursue their career goals. That's ethical leadership.

## ACTION ITEMS

- *Review your responsibilities.* Not knowing your reporting responsibilities is not an effective excuse if you fail to comply with them. Do a bit of research and make sure you are clear on your requirements for reporting issues such as sexual harassment or research misconduct—both what you need to report and to whom you need to report it. You can likely find most of the information you need by searching through your institution's website and online training modules. If you're still unsure, contact your office for faculty affairs or human resources, and they can point you in the right direction.

- *Interrogate your values.* If you've been keeping up with the action items in the previous chapters, you already know the values that you personally and your research group as a whole aspire to embody. You also have a mechanism for asking for feedback! For your own personal values, add a question in your next feedback exercise that first outlines what you value and the type of leader you want to be, and then asks where your actions are aligning with those values and where you need to adjust. For your group values, take a few minutes at the end of your next lab meeting to have a similar discussion.
- *Have the conversation (and write it down).* Talking with your lab members about mental health may feel awkward, but I can almost guarantee that even if you stumble a bit through the conversation, your group members will be grateful that you care enough to bring it up. And I can fully guarantee that you would rather deal with an awkward conversation than the guilt of wondering if you could have done more to help someone. There are many ways to approach the conversation, but the important message to convey is that mental (and physical) health are both more important than any research result, that people can tell you if they are struggling, and that you support their taking the time to get professional help when needed. If you have a lab policy manual, outline these points there as well, so that it's in writing and available when someone most needs it.

# 12

# COMMUNICATION

## You Said It but Did They Hear It?

"HOW MANY OF YOU ARE DOING STRIDES EVERY WEEK?" When only a sparse number of hands go up among our group of amateur runners, our coach's face falls and his chin drops to his chest as he looks at the ground in disappointment. I have a visceral reaction seeing this, because I know exactly how he feels. He thought he had communicated to all of us that we were supposed to be practicing this workout technique, which involves short bursts of ramping up running speed, on a weekly basis to practice good form. However, even though he had probably told us this multiple times and written it into our workouts, the confused look on most of our faces belied our thoughts: "What are strides?" He was communicating it, but we weren't hearing it.

One of the toughest aspects of communication is that we think we've communicated something once we've written it or said it, but in reality, communication happens only when the other person has received and understood our message. Or, in the words of playwright George Bernard Shaw, "The single biggest problem in communication is the illusion that it has taken place." Perhaps you, too, can empathize with my running coach. Whether it's the instructions you gave to a member of your research group for setting up an experiment or ordering supplies, or the article you assigned the students in your class to read before the next lecture, you've likely found yourself in the situation where you thought you had communicated a message but it becomes clear that you are the only one thinking that.

Why is communicating so difficult? One of the clear challenges is the limited ability of our brain to retain the information we hear in a conversation. An often-cited statistic derived by communications scholars Laura Stafford and John Daly is that five minutes after a conversation, we are able to recall only about 10 percent of what was said.[1] Imagine what happens hours or days later. If you don't believe this is true, take a moment to think back to your dinner conversation or the first meeting you had yesterday, and write down how much you remember of what was said. You might exceed 10 percent, but chances are you've still forgotten the majority of the conversation.

Looking back over the previous chapters, the importance of communication is everywhere—your effectiveness as a leader depends on your ability to convey information, whether that is the logistics of how your lab operates, the values and behaviors that you support, or your individual feedback on how each lab member is performing and how they can improve. We've touched on communication skills and strategies throughout the chapters, but here we will bring our focused attention to the topic—this is my chance to practice communicating effectively about how you can communicate effectively.

## Be Transparent

As we saw in the last chapter, the difference between being an ethical person and being an ethical leader is your ability to communicate your values to others. But how do we communicate our values? There will be times when you talk about your values directly—perhaps when leading a session on safety or research ethics. But most of what you convey about your values will come from the way you communicate with others about everyday business, and one factor in this is transparency. In a study of military cadets, transparent communication from leaders was directly linked to the cadets reporting higher levels of engagement in their work.[2] Importantly, the researchers also found that the behavioral integrity of the leaders was a mediating factor. Thus, to reap the positive effects of transparency, leaders had to back it up with ethical behavior.

What does it mean to be transparent? Similar to integrity, there are many definitions for transparency in the leadership literature, and we will use a tangible example—being transparent means that people can see what is on the inside. If you walk through the dairy section of a grocery store, you might see many different brands of milk on the shelves. Some may be in cardboard cartons and others may be in glass bottles. All of the containers should hold the same thing, but some are more transparent than others. This matters because if you can't see the milk inside the carton, you just have to trust that it's there. This may not be a huge leap of faith when it comes to buying milk, but transparency and trust are significant when it comes to the impact that your leadership and decision-making have on the lives and careers of those on your team.

If you've been working through the action items in each chapter, then you are already demonstrating transparency in your leadership by articulating the rules and values of your lab, seeking to uncover the hidden curriculum, and being open about your mistakes and failures. Another area in which transparency is critical is your decision-making. Decisions usually involve change, and change is often frightening or unsettling to others, especially when they don't understand what is happening or why it is being done. Thus, decision-making is one of the most important places for leaders to demonstrate transparency.

Let's revisit a situation we've considered a few times in the previous chapters: a member of your group is working on a project, but they are struggling to make progress. Imagine you have a progress report deadline coming up, and your continued funding may rely on having a certain set of experiments completed by a specific date. What do you do? One option may be to engage another member of the lab who can help in the final push to get everything done before the deadline. While this may be the best course of action, how it plays out among your group members will depend on your transparency (and your communication skills). First, recognize that for the individuals involved, changing roles or adding a new team member may bring about a significant change in their daily plans or in the trajectory of their project. Then, be transparent in communicating with each lab member about why you are making this decision. This is where mutual purpose can be especially helpful—in this situation, you can share that an essential goal for the lab is to maintain your funding so that everyone can continue doing research. This sets you up to explain that certain deadlines need to be met to make that happen, and that you want to achieve that goal without anyone being overworked or burning out from the stress. Even if your lab members don't fully agree with your decision, you will have prevented a significant amount of the frustration, confusion, and mistrust that might have resulted if you had just reassigned the project duties and not explained why.

As a leader, you will also encounter situations in which full transparency is not possible. Perhaps the lab member who is struggling to meet the deadline is also dealing with a health issue that is keeping them from being as productive as usual in lab. You have an obligation to protect the confidentiality of that information. Thus, instead of being fully transparent, the goal is to be as transparent as possible. When you talk with the other lab member who is going to help out on the project, you can explain why their participation would be helpful in meeting the deadline without sharing the reason their colleague can't do the work on their own. The most important thing is to be transparent about your reason for making the decision, and you can do that while still protecting the privacy of your lab members. Additionally, you can recognize that transparency and trust operate in a give-and-take relationship. Demonstrating transparency in your everyday decision-making builds trust, and you can draw from that trust when you have to make a decision and full transparency isn't an option.

## Dissemination versus Dialogue

In the examples throughout this book, we've encountered several different situations in which communication is essential, and they can be divided into two broad categories:

- *Dialogue.* When multiple individuals contribute information and respond to each other's perspectives and ideas.

- *Dissemination.* When one individual broadcasts a message or a set of instructions to another individual or group of people.

Much like the leadership styles we learned about in Chapter 2, neither type of communication is better than the other. They each have their merits, and the key is to deploy the best one in the best way for each situation. Returning to the situation in the lab, where you need to reassign project duties among group members to meet your deadline, you could choose to just disseminate this information by email. But you are likely to achieve a better outcome, both for the individuals involved and for the project, if you instead approach this communication as a dialogue. On the flip side, one mistake I made early in my career was to try to have in-person dialogues about things that could be more effectively disseminated in writing. In the early years of my group, whenever a new member joined, I would launch into a conversation about how to get lab keys, join the email LISTSERV, find a desk and lab bench, access the group-meeting calendar, and on and on. This became exhausting for me and overwhelming for them—it was both inefficient and ineffective. The solution came when we finally drafted our first lab policy manual and I realized that this information could live there and be conveyed by dissemination rather than dialogue. We now have a section in our policy manual entitled "I just joined! Where's my desk?" that has all of the logistical information a new group member needs to get settled into our lab systems. I no longer have to spend a half an hour talking through it all, and they can work through each item at their own pace instead of being overwhelmed by trying to absorb everything from a single conversation.

Since we have already given considerable attention to dialogue skills in previous chapters, we'll focus on dissemination for the remainder of this chapter—the what, when, and how of effectively conveying information so that others will receive and understand it.

## Start and Finish with Listening

Given that dissemination involves broadcasting information to an individual or group of people, it is easy to assume that this way of communicating is a one-way street, but that is not the case. In fact, the most important first step for dissemination involves listening. To communicate effectively, you need to understand what people already know and what they need to learn from your communication. Figuring this out may involve literally asking questions and then listening. When I'm preparing to launch into a new professional development series with my research group, I will often ask what they already know about the topic and what they are hoping to learn, as this helps me to ensure that everyone is starting with a common foundation of knowledge and that the content I provide will be helpful and interesting. Other times, the listening is metaphorical. For example, if the fire alarm is going off and everyone is trying to evacuate, that is clearly not a good time to stop someone in the hallway and start a conversation about whether they understand the escape route. Instead, "listening" means paying attention to body language and movements that might indicate that those you are communicating with need directions or other assistance for getting out of the building. Even when there is no conversation, you are still gathering information about what people need to know so you can communicate what they need to hear.

Dissemination should also end with listening, as this is how you assess whether you have effectively communicated what you were hoping your audience would learn. This doesn't have to be overly complicated—in the example of my running coach, he could finish our preworkout talk by saying, "So, next week everyone is going to finish at least two workouts by doing . . ." and then pausing until we all yell out, "Strides!" Sometimes listening doesn't even require you to ask a question. In my lab, we have disseminated our group's values by emblazoning them on coffee mugs, and every time we have a discussion about recruiting a new lab member, listening to the conversation tells me how effectively everyone has internalized them. When the discussion centers around words like *initiative*, *inclusion*, and *growth mindset*, then I know we're getting it right. And when that is not the case, it presents an opportunity to remind everyone (myself included) about our shared values. As we'll see in the next section, ending by listening not only helps you check the effectiveness of your communication but also has the bonus effect of increasing knowledge retention over time.

As you practice these "listening" skills with your research group, you are likely to see how they translate to other forms of communication as well. For example, when you are invited to give a research seminar or talk, before planning out the content of your presentation you are likely to think about who will be in the audience (and ask these questions of your host if needed). Will your audience be filled with experts in your subfield or individuals with broader expertise across your discipline? Will most of the attendees be undergraduate students, graduate students, or faculty? You can also listen after your presentation by taking note of the questions you receive from the audience. In addition to providing an opportunity to share more information about your work, their questions will show you whether there were points in your talk that your audience missed or misunderstood, and you can try to elaborate on those the next time you give that presentation.

The higher the volume of information your audience receives on a daily basis, the more important your listening skills become in your communications with them. As I mentioned in the introduction, although "leading up" is an area of leadership that we won't delve into in great depth, many of the leadership skills in this book are applicable to it, and this is a particularly important example. When it comes to communicating with your funding agency program officer, department chair, dean, journal editor, or any other person who holds a modicum of control over your work and your future, efficiency is essential to effectiveness. Taking email communication as an example, you may be sending just one email, but if your recipient is in a leadership role, it's likely that your email is one out of hundreds that they receive that day. The effectiveness of your email in producing the desired result will come down to how efficiently you can communicate your request, which in turn requires you to "listen" for cues that reveal what that person already knows and what you need to convey.

If you're not sure how much information they already have and thus how much you need to share, you can use a multitiered approach, where you share information in varying levels of detail and your reader can choose which version works best for them. This approach has become popularized in informal communications by the comment "TL;DR," for "too long; didn't read," and offers the recipient a concise summary in addition to a very detailed message. A particularly effective professional example came from my kids' school district during the COVID-19 pandemic. Knowing that some parents wanted detailed information but others were just trying to stay afloat while balancing work and home schooling, the district used a 10-100-1,000 format in its communication with families, in which they shared new information in messages of 10 words or less, 100 words or less, and more than 1,000 words. A 10-word version, for example, might have been "School won't resume in person for at least three weeks," and the longer versions would include information on why this decision was being made and how the logistics of returning textbooks, delivering art supplies, and providing school lunches would function during that time. Taking a lesson from this format, when I find that I need to write a long email to a busy person, I try to write a one-sentence summary at the start, and I use bold text for any actions or responses that are needed on their part. I then "listen" to their response (or lack of response) to gauge how effective I was in communicating with that individual and what I might want to do differently the next time.

## Set the Stage for Learning

One of the tough realities of communication is that no matter how effective you are, almost nobody is going to remember 100 percent of what you tried to convey. Or if they do, they won't remember it for very long. If you teach classes and have used some of the best practices for student learning, you're already familiar with how to help people prioritize and retain information.[3] You can apply these same principles to communications with your research lab.

We'll start with something that may not be entirely intuitive but can have a huge impact—reducing stress. Research has shown that stress decreases the level of recall after a conversation.[4] You've likely experienced this yourself. Ever had a day when you had a huge grant deadline looming and a conflict playing out in your lab or personal life, and then you had to attend a faculty meeting? There's a good chance that your stress level was up and your memory of the meeting's content was much spottier than usual. The same thing applies to your lab members. You may not be able to remove all of the stressors that they are facing, and it's not your responsibility to do so. It is your responsibility, though, to create a healthy lab culture where everyone is supported and feels a sense of belonging, and this can in turn reduce stress levels and help each person to function at their best.[5] You can also think about intentional practices to reduce stress. Some faculty take a "mindfulness minute" at the start of their group meetings to encourage everyone to relax and focus on the task at hand. Organizing regular social events for your lab can also be a great way to promote well-being and help each person show up to work ready to learn.

Once you've set the stage for effective communication by creating a healthy environment, it's time to think about how you deliver content. One of the most important practices you can employ is to lead with *why*. Before you start sharing information, explain why that information is going to be important to your group members—perhaps it will help them with finding a job in the future or it's critical for their safety in lab. Just as being told about the "real-world" application of a scientific concept increases knowledge retention among students in your classes, knowing how the information you are communicating will be useful can improve your group members' retention of that knowledge.[6] I've found this to be particularly important when I deliver my annual "state of the group" talk on the first day of our retreat. Much of this talk is focused on topics such as lab finances and proposal due dates, things that I think about on an almost daily basis but that are far from the top of my lab members' minds. Before diving into the presentation, I take a moment to explain that while the immediate relevance of some of the information may not be obvious, my sharing it is important for the success of our group. Specifically, I elaborate on how this annual talk contributes to the culture of transparency we want to have in our lab, and how it provides important background and motivation for the work we will do throughout the remaining days of the retreat.

Another helpful approach for making information "stick" is to take advantage of both verbal and visual communication, especially when it includes pictures and diagrams.[7] Here, design matters. When it comes to especially important information, such as your lab values or safety rules, work together with your group to create graphics that will make the information both understandable and memorable. The group approach has the added bonus of engaging your lab members in active learning, which can also improve recall of important information over time.[8] And active learning doesn't have to stop there—as you communicate, think about all of the opportunities you have to promote participation. Earlier, we discussed working together with your group to create a policy manual, and how this helps ensure that each person actually knows what your group's policies are. You can also build participation into activities that are typically one-way communications, for example by collaborating to create the group meeting schedule or working through case studies together to discuss research ethics. Finally, research has shown that a change of environment can increase memory.[9] While it may be convenient to hold your group meeting in the same conference room every week, moving it around occasionally can be beneficial. This also underscores why a group retreat can be so special—whether the "away" location is two floors up from your lab or a two-hour drive from home, the change of scenery alone can change how we process and store information.

## Say It Again

Approaches like these can dramatically increase the amount of information that people retain, but no matter how effective you are as a communicator, there is a good chance you're going to find yourself at some point looking out at a sea of confused faces because nobody remembers hearing the thing that you thought you had clearly communicated. This isn't anyone's fault. Our brains are wired to be able to forget things, and this is actually a critical part of how memory and creativity work. Think about what your living space would look like if you never threw anything away. The rooms would become so disorganized and full that it would be tough to move around or to find the things that you need. Just as we throw things away to keep our living space tidy, psychologists propose that we toss out unnecessary or distracting memories by forgetting them.[10] Unfortunately, we sometimes also forget important information. This means that, as a leader, you are going to need to practice the art of overcommunication.

Overcommunication is exactly what it sounds like—communicating something more times than it seems like you should have to. I used to think that doing this was a communication misstep, as I was being redundant and inefficient. But I've come to realize that this is rarely the case. You'll know when you've gone too far and over-overcommunicated. If you find yourself sharing a piece of information with your group for the fifteenth time and you notice people's eyes are starting to glaze over, or they're finishing your sentences for you (which is still good active learning!), then you've communicated sufficiently and don't need to say it again—at least for a little while.

Just as important as saying something multiple times is saying it in multiple ways. Although the theory of "learning styles" has consistently failed to stand up to scientific inquiry, each person in your lab is still unique and has their own system for absorbing, processing, and retaining information.[11] For example, when it comes to group meeting times and formats, one individual may prefer to have all of the information on a piece of paper that hangs on their office door, whereas someone else is more likely to remember the meetings if each event and its relevant information is entered into their online calendar. For really important topics, such as safety, there is likely some mandatory training, and you can build in additional communication to reinforce critical information. This could take the form of a group safety manual that everyone reads and signs on a yearly basis, short discussions you initiate on safety topics at the start of each group meeting, or signs and posters that hang in your lab as reminders of key safety practices. Communication style is an area where you can think creatively and break from the norm. When my group generated our word cloud of shared values, we wanted to communicate this in a way that would tell others what we value and consistently remind us, as well. That's how we ended up with coffee mugs.

## Communication Squared

Of all of the topics in communication, we've left what is arguably the most important for last—communicating about how you will communicate. We live in a world where information is coming at us rapidly and from every angle. Within a five-minute span, you can check your email, receive a text message, read a poster in the hallway, accept a calendar invite, and be tagged in a post on social media. As we just discussed, it can be helpful to harness parallel channels for communicating information to your group and others, but it's important to be clear in advance about what these channels are and how each one will be used. Our lab keeps our policy manual in a folder on our shared drive, where it's readily available anytime someone needs it. Of course, if we don't tell new members where to find it, then the chances that they will stumble upon it in that corner of the shared drive are fairly small. Similarly, if important group announcements are made on a group messaging platform but someone is expecting to see them only in their email, they may have a massive list of notifications on an app they've never opened and wonder why they didn't know about the important meeting that morning. In our lab, we manage this by keeping a list in our policy manual of the different communication platforms we use for group messages, equipment calendars, and other important functions. This way, when a new member joins the lab, we just need to make sure they know how to find the policy manual, where all of the other information on communication systems can be found.

Communication is a two-way street, so it's also critical to make sure your group knows how to communicate with you. There is no single best way to manage this type of communication, and what works best for you and your group will likely change over time. In part, this is because technology is constantly changing—we now have access to programs and apps that simply didn't exist a decade ago. Also, as you move along your career trajectory, your work life will change. Early in your career, email may be an effective way for group members to get your attention. Later, as you take on more responsibilities, you may find yourself wading through more than a hundred emails a day and missing important information from your lab. For this reason, a good general practice may be to have a separate channel of communication that the group can use—typically a group messaging app. I've heard many faculty members share how useful they find this approach. It's also helpful to have different communication channels for different levels of urgency. For example, my group members all have my cell phone number and know that they can call or text me if something comes up that is urgent. Alternatively, because I do keep a close eye on my email inbox throughout the day, I've let them know that if a message is particularly timely, they should start the subject line with "TODAY" or "URGENT" and I will make sure to prioritize those messages.

In addition to knowing how to communicate information to you, it's important that your group members know how to find time to meet with you. An eye-opening moment for me was the first time my group gave me feedback and asked me to be better about communicating my travel dates. I have an open-door policy, and they know they can stop by and talk anytime I'm in my office, but what I didn't realize was that when my door was closed they had no way to know whether I was in a meeting for an hour or out for the entire week. Someone in my lab pointed out how frustrating it was to stop by my office several times in a day, hoping to find me, only to later learn I was away at a conference. I agreed—that is frustrating. As a result of this feedback, I created a calendar where I share my travel dates with the lab. A few years later, I went a step further and started sharing my Outlook calendar with all of my group members so they can see what times I'm busy and when I'm free. Now if they need to meet with me, they can look for a free time in my schedule and send me a calendar invite.

As our lab size grows and shrinks and my schedule gets more or less busy, we are constantly adapting our communication and scheduling practices. We can do that and keep everyone aware of the current practices so long as we communicate openly about what is working, what is not, and what we've changed.

## ACTION ITEMS

- *Listen and learn.* Think about something that you've been trying to communicate to your group but you feel isn't getting through. Perhaps it is about the format for group meeting presentations or how to manage drafts of a manuscript. Take a step back and listen—create a short survey or have a group discussion to ask: What do you think our policy or practice is? What questions do you have about this? What do you want to learn?
- *Follow the evidence.* Choose something that you need to communicate to your group in the upcoming week—perhaps a

new group meeting schedule, a safety policy, or advice on making research posters. Implement two approaches from evidence-based pedagogy (for example, active learning and using pictures and diagrams) in your communication strategy, to make the information more likely to stick.

- *Communicate about your communication.* Review the ways that you communicate with your group and how you expect them to communicate with you. Is all of this written down somewhere that everyone can access? Does everyone know where to find that document? Does your group seem to know how these communications work, or have there been gaps lately, such as several people showing up at the wrong room for a meeting after a location change was communicated? Revisit and refine your description of these communication practices . . . and then communicate that to your group!

# III

# *COACHING FUTURE LEADERS*

# 13

# PASS IT ON

## Cultivating the Next Generation of Leaders

THINK ABOUT HOW OFTEN YOU'VE ENCOUNTERED a situation in your faculty career and thought, "I wish someone had prepared me for this!" Now is your chance to do exactly that for someone else.

Up to this point, we've been focusing on how to best lead yourself and others, and if you were leading a team of people who would forever be under your management, then we could stop here. That's not the case, of course. If you're leading a research group comprised of students and postdocs, then part of your job is training future leaders.

I'll never forget the moment when the first PhD students in my lab were getting close to graduation, and I was struck by the reality that I would be the only permanent member of my group. Those first students had been there since the start, unpacking boxes and figuring out protocols, and I had subconsciously let myself believe that they would always be there. At that time, it was heartbreaking to think that the group I lead will forever rotate and change, with few people staying longer than five years. Several graduations later, I now embrace this aspect of faculty life, because that constant rotation means continuously getting to welcome new people into the group, along with their enthusiasm, innovative ideas, and fresh perspectives. Even more exciting, it means seeing each of those individuals grow and develop as researchers and leaders and then go out into the world and do incredible things. Now one of my favorite activities is connecting with former lab members to catch up on the latest news about their lives and careers, and I've come to realize that they never really leave the group but instead become part of an ever-growing network of lab alumni.

Some of the people you currently lead may stay in academia and become faculty. In that case, teaching them the leadership skills you've learned will be directly applicable to their future career. However, these skills are just as important for those individuals—probably the majority of your group—who will pursue careers beyond academia, whether in industry, policy, entrepreneurship, science communication, or elsewhere. They, too, are likely to find themselves in roles where they are managing their own time and managing other people, and that require most of the same leadership skills we've been discussing. No matter where each member of your research group ends up in the future, you can have a powerful impact on their career by helping them develop those skills now.

You may hesitate at the thought of teaching leadership skills to your group members, especially if you feel like you are only starting to develop them yourself. It's natural to feel that way. Most of what you teach to those in your lab is something you're already an expert in—you've earned degrees and done years of research in a specific field, and that is what has equipped you to mentor others in that area. But you don't need to be an expert to have an impact as a teacher. If you've ever assigned group work or employed the "think-pair-share" method of active learning in a class, you've seen this in action. Just as students can benefit themselves and others by teaching something that they are just starting to learn, you can do the same in your lab.[1] And, as we'll see, you don't have to go it alone. In addition to the resources I highlight in this book, you can find expert leadership advice in many other books, articles, and podcasts and use it to grow and learn together with your group members.

## Mindset Still Matters

Viewing the development of new leaders as part of your job may require a shift in your mindset, and we can get to the root of that by thinking back to how we saw ourselves as graduate students or postdocs. Looking back to those times in your life, how would you have described your work and your goals? Did you see your sole aim as getting research done so you could publish papers, or did you also have hopes of building a skill set that would prepare you for your future career? Here's where a critical gap may appear—at the time, you may have thought that doing research and publishing papers was all of the training you needed to pursue a faculty career. You may now find that notion laughable, and this is where the shift in mindset is needed. If you are leading a lab, you probably already recognize that the students and postdocs who work with you are hoping to gain the skills and expertise they need to succeed in their next job. The challenge is that academic culture continues to broadcast the message that success is only measured in research output. To push back on this deeply ingrained bias, take a moment to think about the most senior members of your group, with whom you have had the most conversations about future career plans. Envision each of them in the role that they aspire to—a senior scientist in industry, a policy advisor meeting with a congressional representative, an outreach coordinator organizing a citywide science festival, a faculty member leading a lab. Now think about the skills they will need to succeed in those roles and whether they are gaining them from their experience in your lab.

If you're like me, the outcome of that exercise feels daunting—almost overwhelming. As we work through these last chapters of the book, I hope you will see that equipping future leaders often just requires small changes to the way you lead your research group, and that you are already doing more than you think. But before we get to that, we're going to talk about another mindset that still matters: how you view the capacity of your group members to be future leaders. Remember the Pygmalion effect mentioned in Chapter 9? Robert Rosenthal and Lenore Jacobson found that teachers' perceptions of a student's capacity for growth had an impact on the actual growth of that student.[2] This can also apply to leadership development. In a study of leadership training, Diane Wood Allen, then vice president of operations and chief nursing officer at Concord Hospital in New Hampshire, surveyed a group of nurses to understand what had a positive impact on their development as leaders. Allen found that receiving support and encouragement from mentors was an important factor, and that the nurses' confidence influenced their likelihood of taking on leadership roles.[3] Similarly, how you interact with each member of your group sends a powerful message about their potential as leaders.

## It's Already Happening

Just as your mindset is already shaping the way that you mentor those around you toward leadership, your actions have a similar effect. I'll never forget what happened one day when I was getting ready to go to the grocery store and I told my then preschool-age child that it was time to get ready. He was playing with an old laptop computer, and he looked up and said, "Okay, just let me finish writing a few emails first." (Of course, he didn't have an email account and the computer wasn't even turned on.) That was the day I recognized that the behavior I model has much more impact on my children than anything I might tell them about how to behave. Research on parental modeling shows that children learn from and have an increased likelihood of replicating their parents' actions across a wide range of behaviors, from exercise to alcohol use.[4] This principle also extends into the workplace, as reflected by business ethics experts Kevin Brown and Linda Treviño's finding that individuals who have models of ethical leadership are more likely to be ethical leaders themselves.[5]

Whether intentionally or not, you are training future leaders every day through the words and behaviors you model—the question is whether you are happy with the training you are providing. The next time you wrap up a meeting with members of your group, take a few minutes alone afterward to reflect on your behavior in that meeting and think about the model you are creating. If the individuals in your group were to mirror your behavior as leaders of their own research team five or ten years from now, is that something you would be proud of or embarrassed by? Just as in parenting, we'll never be perfect leaders and we'll all have some bad days, but what's important is to keep growing and improving so that most of our behaviors are something we would be proud to see others emulate.

## From Modeling to Teaching

While your own actions as a leader clearly influence your training of future leaders, it doesn't have to stop there. If you were teaching someone in your lab how to perform a specific technique—let's say a polymerase chain reaction (PCR) experiment—imagine if you had them just watch as you silently moved around the lab, gathering up tubes of reagents, pipetting them into other tubes, and programming a lot of numbers into an instrument. Would they be able to replicate what you'd just done? We already know that if we're going to teach someone a new technique and expect them to learn it and eventually perform it on their own, we need to not only model with our actions but also explain with our words. In the case of the PCR experiment, you would describe each step of the process by talking about what is in each tube that you pull from the freezer, how much of each reagent you are using, and how to program the thermocycler instrument into which you will ultimately put the samples. In the same way, you can be intentional in sharing the knowledge and skills you've gained in the preceding chapters of this book with your group members, so that they will be equipped to replicate effective leadership in the roles they hold now and when they are on their own in their future careers.

The thought of teaching others about leadership might feel scary or awkward, but it doesn't have to be that way. If you've ever taught someone a new research technique, you can use the same approach to teach a new leadership technique. Think back over the skills we've covered in the first two sections of this book—time management, conflict resolution, giving and receiving feedback, and more. The knowledge you've gained by learning and putting these skills into practice is something that you can teach to your own group members. And you can use the same events and opportunities that you've set up to convey research information to also teach leadership skills.

When it comes to mentoring your lab members on research, you probably follow at least two different formats—mass dissemination at a lab or subgroup meeting, and individual discussions that take place in person or by email. Each of these serves its own purpose. When you share information in a group meeting, it is usually information that you know multiple people are going to need, whereas individual conversations tend to be focused on specific examples or questions that are unique to that person and their goals. You can leverage these same formats in your dissemination of leadership information.

Full disclosure: this is something I learned by realizing how badly I was failing at it. Early in my faculty career, I had a few leadership tips that I would regularly share with group members, such as how to cope with failure or how to manage time to organize the daily tasks of a project. However, I was sharing these only in the context of individual interactions. Over time, I would find myself having a conversation with someone in my office and starting to launch into my leadership lesson, only to stop myself and ask, "Have I already told you this?" It hit me that I was probably giving the same information to some people multiple times, while others were never hearing it. My system for disseminating leadership lessons was both inefficient and inequitable.

As a result, I decided to try something new—I put myself on the group meeting calendar as a regular presenter to share instruction on topics related to professional development. Our group meetings still have the typical presentations about current literature and research progress, and we now also have a fifteen- to twenty-minute presentation on a topic that extends beyond the research. Instead of sharing leadership advice only in one-on-one conversations and having to repeat it over and over, I can share with everyone in the group meeting. Even better, I can keep the slides from these presentations on our shared drive so that the advice continues to be accessible to everyone. On a practical note, I've found that rather than tackling a different topic each week and trying to cover all that I want to in the limited time I have, it is easier to choose a topic—such as time management or goal setting—and break it into a multipart series that extends over several meetings. As with the chapters of this book, I try to finish each presentation with an action item that requires the members of my lab to take a few minutes to do or plan something concrete that will have an immediate impact on the way they lead themselves or others. In an unexpected twist, the members of my group also started volunteering to give presentations, and we have had student-led professional development talks on a range of topics, including how to maintain an Individualized Development Plan, tips for negotiating a job offer in industry, and how to make visually appealing and informative figures for presentations and manuscripts. Importantly, I still have a chance to talk about leadership during individual meetings, but now those conversations can be more focused and effective. Instead of rehashing my general advice on a topic, I can refer to a previous group meeting presentation for the basics and focus the conversation on more nuanced points that are specific to the situation at hand.

You may still feel unprepared to teach others the leadership skills you are just learning and developing for yourself. I encourage you to lean into that discomfort, as there are two hidden benefits to this approach. The first is that teaching is a great way to learn.[6] You can read a chapter in this book and take away some knowledge about leadership, but having to synthesize that knowledge and explain it to others naturally creates a much deeper level of understanding. Second, talking about leadership gives you an opportunity to compensate for the instances when you haven't gotten it quite right in the leadership that you model. We can turn those mistakes into teachable moments when we share them with others and explain what we wish we had done differently.

## Call In the Experts

A final point to remember as we transition from developing our own leadership skills to teaching others is that being a teacher doesn't mean that you have all of the answers. You may be used to teaching classes on a topic in which you are an expert. You have studied and gained experience with the material over years and possibly decades. Nevertheless, there is always more to learn. I have personally found that no matter how many times I teach a class or how well I know the subject area, I almost always get a few questions from students to which I don't know the answers and have to say, "That's a very interesting question—let me read up on it and get back to you next class session." In those instances, I dive into the literature or consult a colleague to find an answer that I can bring back to my class. The point is that if it's okay to rely on help from others when teaching something that we're an expert in, then it's *definitely* okay to do this with a topic we're still learning about.

Sometimes seeking help in teaching leadership skills can mean literally calling in an expert for the equivalent of a guest lecture. The universities where I've worked have offered leadership development programs through their human resources departments, and our research group has taken advantage of these offerings to participate in personality assessment activities, using the tools we discussed in Chapter 2, and in day-long workshops on conflict resolution and other topics. If your institution provides something similar, it may not be completely free, but professional development is often heavily subsidized to make it financially feasible. And if your institution doesn't offer these resources at all, you can still benefit from professional leadership advice at a reasonable cost. Consider choosing a book on a leadership topic. (I've referenced many great options throughout this volume, and there are many more out there.) Buy a copy for each of your lab members, and then meet every few weeks as a group to discuss a chapter. Or find a relevant TED talk or other leadership video to watch together and discuss. No matter what amount of time and money you are able to commit, there is a way to tap into the leadership expertise of others, and doing so as a group will maximize the impact and also help your lab grow closer as a team.

In addition to these planned opportunities for sharing leadership knowledge, you can look for more spontaneous opportunities to pass along advice. In the same way that you or a member of your group would share a new research paper that could be of interest to others in the lab, you can also share when you have come across an article or a podcast on leadership that you think others in the group would benefit from. This kind of information sharing can come into play in your individual interactions with group members as well. In Chapter 15 we'll talk about mentoring people to lead through challenging situations, and that is definitely an opportunity to share wisdom and resources from leadership experts.

Just as you weren't trained to be a leader, you weren't trained in developing other leaders. Nevertheless, you can have a powerful, positive impact if you work at it. Imagine what it could look like when someone you currently mentor is on their own as a group leader and they encounter a situation that requires an important leadership skill—whether it's giving candid feedback to someone on their team, organizing their calendar to get a proposal finished on time, or mediating a conflict over project assignments. Now imagine that when they encounter that situation they think, "No problem. My training prepared me for this!" You can make that a reality.

## ACTION ITEMS

- *Find the right mindset.* Think back to your time in graduate school and reflect on how much you have grown as a leader since then. How did you manage projects, your time, or interpersonal conflict back then, and how have you learned to manage these better now? Recognize that each of your lab members is capable of the same growth and that you can play a part in that. Before your next meeting with a lab member, pause to envision them ten years from now in a future leadership role—whether in academia, industry, government, or elsewhere. Picture them leading a team, and then approach your meeting with that individual with the goal of helping to prepare them for that role.
- *Check the model.* Reflect on three interactions in the past twenty-four hours when you were in a leadership capacity and someone was watching you—for example, when you were communicating information, running a meeting, or making decisions. In what ways did your behavior align with the type of leader you want to be, and in what ways did it not? Choose one of the times when you did not model what you wanted to, and plan how you will turn this into a teachable moment for your lab. (If you need some help doing this, you can read ahead into the next chapter.)

- *Teach!* Choose a topic from the first twelve chapters of this book and make a plan for how you will teach that information to your lab members during a group meeting or other team event. You can share your own wisdom, insights, and experiences and also call on experts by including an article on the topic or a brief video that you've found to be helpful. End your lesson with an action item so everyone can start to put their new leadership skill into practice.

# 14

# VULNERABILITY

## Nobody Is Perfect and That's a Good Thing

WHO DO YOU RESPECT? Who are the people you look up to most, and perhaps even hope to emulate in your own life and career? Have you ever stopped to think about *why* you respect those people? I'm sure there are many reasons, but I'm guessing that one reason you won't find on that list is "because they are perfect and never struggle or make mistakes." Yet one of the barriers to admitting our own mistakes and failures is the fear that other people will respect us less.

We saw in Chapter 6 that the ability of leaders to admit their own shortfalls is a key ingredient in building psychological safety within their team. It can also be an important component in building respect and relationships with those you lead, as well as in helping those individuals develop into leaders themselves. Management expert Richard

Farson and author Ralph Keyes highlight a number of examples as they discuss the power of being a "failure-tolerant leader."[1] In one example, a former executive at the aviation giant Lockheed describes hearing then-CEO Dan Haughton describe turning down the opportunity to purchase competitor Douglas Aircraft, a decision that ended up being a financially costly mistake. The executive said that rather than reducing the respect he had for Haughton, hearing this story had the opposite effect, as it helped him see that if it was okay for their leader to make mistakes, it was okay for them to do so as well.

That last point is especially critical. Being a leader means venturing into uncharted territory and making decisions even when you're unsure of the outcome. Under those conditions, it's impossible to always get it right. As you think about helping each member of your research group develop as a leader, one of the greatest gifts that you can give them is the permission to make a mistake or to fail. In Chapter 13, we discussed the value of sharing leadership best practices and talking through examples of successes. We can add depth and power to those discussions by thoughtfully layering in stories of our own struggles and uncertainties, both to help others avoid the pitfalls that we've encountered and to show that being a good leader doesn't have to mean being a perfect leader.

In this chapter, we will explore the use of vulnerability as a leadership development strategy. But first, I want to take a moment to recognize that leadership itself is inherently vulnerable. Social work scholar and author Brené Brown defines vulnerability as "uncertainty, risk, and emotional exposure," and it's difficult to imagine being a good leader—especially in a research context—without those three conditions being regular companions in your day-to-day work.[2] The more effectively you can prepare the members of your team for this reality, the more potential they will have to thrive in their own future leadership roles.

## Make It Attainable

While we typically think of role models as having only a positive impact on those who look up to them, a study by leadership expert Crystal Hoyt and her team shows that role models can either build or deflate our confidence and have a positive or negative impact on our mental health. What makes the difference is mindset.[3] Hoyt and colleagues surveyed a group of individuals to assess whether they had a fixed mindset, believing that "leaders are born," or a growth mindset, thinking that "leaders are made." They then primed each individual by asking them to think about one of their role models before asking them to perform a leadership task. When assessing the participants' confidence and affect during that task, the researchers found that the growth mindset group had both higher confidence and a lower anxious-depressed affect than the fixed mindset group. They also performed better at the task. Even more important, the authors explored the impact of changing the participants' mindsets. They found that intervening to promote a growth mindset could produce similar benefits of confidence and well-being in subsequent leadership situations.

What does this mean for you as a group leader? Even if the members of your lab don't aspire to follow your career path, they probably still look to you as a role model. When we make it look like leadership is "easy" or "just comes naturally" to us, we can make it look unattainable to others and thereby have a negative impact on their confidence. At the other extreme, if we make it look miserable or impossible, we also don't do much to inspire others. It's when we can be vulnerable in sharing our struggles and talking about how we have overcome them and grown that we most effectively help those around us to see and believe in their own potential for greatness as leaders.

If you're like me, you have no shortage of leadership struggles to choose from when sharing with your group. We'll talk later about how to create healthy boundaries in what you share, and clearly there are also situations in which confidentiality concerns will prevent you from being fully candid. Nevertheless, I'm sure that will still leave quite a few times when you communicated poorly, overbooked your calendar and missed a deadline, shied away from giving candid feedback, or weren't sure how to tell the group about a rejected manuscript or grant proposal. If you're not sure where to start, you can always go back to where we started this book—when you felt that sense of "I wasn't trained for this." Unless you say otherwise, your group probably assumes that you came into your job with all of the skills and knowledge that you currently possess, especially if they are relatively new to the lab and haven't had the chance to see you developing as a leader. Whether you seize the opportunity during a group lunch or plan it out as part of a professional development presentation, you can share with your group how you struggled with the realization that you had stepped into a job you weren't prepared for, and highlight some of the mistakes you made as you scrambled to develop the skills you needed. And your story doesn't have to end with struggle or failure. You can also talk about how you recognized the challenge and overcame it (or are still doing so). You might point to the first time someone shared leadership advice with you and how it allowed you to grow and approach a situation in a better way the next time. When those around you can see your growth, it can give them the confidence to see the same potential for growth in themselves and equip them with the wisdom they will need when they encounter similar challenges in the future.

## More Than Mistakes

In communicating with vulnerability, sharing mistakes and failures is an important place to start, but it doesn't have to end there. A healthy approach to vulnerability also includes talking about the struggles that can underlie our work every day, such as self-doubt, feelings of impostor syndrome, and burnout. Human resource development expert Holly Hutchins points out that academia is almost perfectly set up to cultivate impostor syndrome, as it attracts high-achieving individuals and places them in a competitive environment where the metrics of success are vague and emotional support is often lacking.[4] Sound familiar? I'm guessing you can relate. Unsurprisingly, studies have shown that impostor syndrome is prevalent among people across multiple domains of academia, from undergraduates to faculty.[5] Interestingly, Hutchins also finds that mentors can play a mediating role to reduce impostor syndrome by influencing whether individuals attribute their successes and failures to internal factors, such as ability, or external factors, such as luck. And while rigorous studies on the treatment of impostor syndrome remain lacking, experts do suggest that openly discussing self-doubt and failure in a healthy and supportive context can have a therapeutic effect.[6]

Talking about our struggles with impostor syndrome and other psychological barriers can be more challenging than sharing about our mistakes, as a story about a mistake usually has a conclusion, but the other types of struggles are often ongoing. As researchers, we're in the business of problem-solving, and these psychological struggles typically represent problems we haven't yet solved. On the other hand, that can arguably make these examples of vulnerability even more powerful to those who look up to us as leaders.

A few years ago, I was giving a professional development presentation on the topic of motivation as part of our group meeting, and I led into a piece of advice by saying, "When I'm struggling to get motivated . . . " Before I could finish the sentence, I was interrupted by a postdoctoral researcher who said, "Wait! You're saying that *you* sometimes struggle with motivation?" The entire room erupted in laughter. People who know me know that I am a high-energy person, and I've received the feedback that those around me view my drive as a major strength. While the contrast between my confession and the impression that others have of me provided an unexpected moment of levity, it also taught me a crucial lesson: the most important struggles we can share may be those in areas where people perceive us to have the greatest strength.

Along similar lines, when we talk about our mistakes or failures, we can include how those made us feel or how they impacted our confidence as leaders. The story of my first grant rejection that I recounted in Chapter 5 is one that I regularly share with the members of my lab, especially when they are also facing a painful rejection. The important message from this example is that failure doesn't have to be the end of the story, and perhaps even more valuable is how the failure affected me as a leader. When I describe my instinctive response to start searching through the job ads to find a new career, I'm revealing that my failure not only stung in the moment but also made me doubt my ability to do my job. If someone in my lab ever finds themselves facing similar self-doubt, my goal is to help them feel a bit less alone and to show them that questioning your abilities is a natural part of leadership and doesn't mean that you're not cut out for the role. Additionally, this story provides an opportunity to talk about my concern that the failure would affect the confidence the group had in my leadership and our research program. As future leaders, my group members need to know not only how to cope with failure themselves, but also how to lead their own teams through that experience.

## Vulnerability Is a Privilege

As we dive into the benefits of admitting our mistakes and sharing our struggles, we need to acknowledge the relationship between vulnerability and privilege. Thinking back to the example of Lockheed CEO Dan Haughton, it is imperative to question whether the response to his mistake would have been the same if he had not benefited from being someone who looks like the majority of leaders in the country. This advantage is further reinforced by the rest of the story—that Haughton remained highly respected in the industry, even after he stepped down as CEO amid massive financial and legal crises.[7]

A 2010 study that sought to quantify this privilege effect found that women in male-dominated roles—police chief, CEO, chief judge—saw their status decrease significantly relative to that of their male counterparts when they made a mistake.[8] Interestingly, the authors observed a similar effect for men in a female-dominated role, that of president of a women's college. The authors concluded that for individuals in what they classify as gender-congruent roles, competence is assumed. However, for those in gender-incongruent roles—meaning that people don't expect to see someone with their identity in the position—their status is much more fragile, such that small mistakes can easily call their competence into question. In a parallel study, leadership scholar Ashleigh Shelby Rosette and social psychologist Robert Livingston found that Black women face "double jeopardy" and are penalized to an even greater extent for their mistakes, relative to Black men or white women, because they have two identities that are viewed as atypical for leaders.[9] Moreover, these studies and examples take into consideration only the gender-binary identities of female and male. Research on the experiences of nonbinary or genderqueer individuals in leadership remains sparse, despite the unique challenges that these individuals can face.[10] A unifying theme for those with marginalized identities is that even in situations where someone won't actually be treated differently for admitting their mistakes, factors such as stereotype threat can make them feel that they will be judged more harshly, making vulnerability more costly for them to put into practice.[11]

The double-jeopardy study from Rosette and Livingston underscores the importance of considering the intersectionality of identities, as it can impact the level of safety individuals feel when sharing their experiences. One illustration of this effect comes from a self-study by three women who participated in a women's leadership program aimed at preparing female faculty and staff for roles in academic leadership.[12] In discussing their experiences in the program, a theme that emerged from the authors was the need to consider intersectionality. Even in a space designed for those who shared a gender identity that has historically been marginalized in academia, the lack of consideration for other identities, such as ethnicity, created a different experience for women of color compared with white women. One of the authors who identifies as a woman of color described an instance when a senior male administrator made a comment about her appearance that made her feel a lack of acceptance as a leader. She added that while she was comfortable recounting this experience to her coauthors, she had not felt safe enough to do so among the larger cohort of primarily white women in the leadership program.

While the ultimate goal is to create an inclusive and just academic system where everyone is represented and enjoys the same privileges, until that is the case, the uneven distribution of privilege will affect how you demonstrate and talk about vulnerability in your research group. As a faculty member, when you have identities that don't align with the majority of other faculty in your field or at your institution, it can change how you talk about your mistakes and failures with your group. If you have identities that make you more likely to have your competence questioned, then you might choose to talk only about instances when your initial challenge or mistake turned out to be beneficial, or about struggles that you have already overcome. If some members of your lab share your identities, then you may opt to be more candid with that smaller group, for whom you serve as an especially powerful role model. As we'll touch on in the next section, vulnerability should always be a choice and not something that you are forced to engage in. Identity will also impact the advice that you give to those in your lab as they think about becoming leaders in the future. They will enter into those positions with varying levels of privilege and varying assumptions about the level of their competence, and you can play an important role by preparing them to navigate how and how much they can demonstrate vulnerability with their own future team.

## The Why Is What Matters

A common misconception about vulnerability is that it means spilling all of your secrets or sensitive information to every person you meet. This is not the case. Vulnerability means that the things you share are sensitive, not that you share everything that is sensitive. When employed well, displaying vulnerability in leadership is a calculated risk that we intentionally take because we recognize its ability to benefit the people on our team.

In her book *Dare to Lead*, Brené Brown highlights that true vulnerability requires both boundaries and intention.[13] She stresses that vulnerability is not directly related to the amount of information that is disclosed, but to the potential of the information to be productive. Brown provides an example of a team leader in an organization that is undergoing significant change, which is creating anxiety among the team. She explains that a simple statement from the leader, acknowledging that they are also feeling anxious about the changes, can be very powerful. Notably, a key determinant of whether vulnerability will be productive is the team leader's reason for sharing it. Brown is clear that sharing in order to gain sympathy or to manipulate the opinions of others is not true vulnerability. Thus, if the leader is just hoping to get the team to focus their blame and anxiety on someone else, sharing their fears will probably not have a positive effect. In contrast, if their goal is to convey to the team that they understand the challenges, then sharing can help team members feel reassured that their leader is committed to working through the uncertainty together.

Sometimes the dividing line between healthy and unhealthy sharing is clear. For example, if a member of your group is questioning their ability to be a researcher because of a mistake they made in setting up an experiment, telling them about a similar mistake you made earlier in your career can help normalize mistakes as part of the research process and assure them that they do still belong and can succeed—that is healthy and effective vulnerability. On the other end of the spectrum, it would clearly be unhealthy and ineffective to storm into your group meeting and spill all of the details of an argument you had with your significant other the evening before—that is venting, not vulnerability, and it is aimed at the wrong audience.

While these situations represent extremes where the decision about whether or not to share information is clear, many situations are more nuanced and ambiguous. Thinking back to the example of my first rejected grant proposal, telling that story in retrospect, after I've earned tenure and established some level of success in obtaining research grants, carries relatively little risk. Deciding to talk about the failure at the time that it happened was a more difficult decision. On one hand, I was aiming to build a lab culture in which every member viewed themselves as a full stakeholder in our research enterprise and we could openly discuss our failures. On the other hand, we all recognized that the stakes were high—if we didn't get funded, I would not get tenure, and that would have a disruptive and negative impact on the members of my lab. They were putting a tremendous amount of trust in my ability to succeed, and I was wary of disclosing information that might undermine that trust. In the end, the "why" convinced me to tell the group about the failed proposal, as building our group culture was worth the risk and I would be sharing for the benefit of others, rather than to gain pity or to vent my frustration. The why also allowed me to approach the discussion in a way that minimized the risk of damaging my relationship with my group. Instead of conveying anger at the funding agency or hopelessness about the future, I was able to say that this individual failure was frustrating but that it did not undermine my confidence in our ability to succeed, and that the plan was for us to learn what we could from this experience and then look to the future and keep trying.

## Remember Your Win

Up until now, we've talked about situations when we choose vulnerability, but sometimes vulnerability is chosen for us. That was certainly the case for me on the day of my first tenure vote. It didn't go well, and I had no choice but to confront a very risky, uncertain, and emotionally exposed future. I'm sure I didn't handle it perfectly—there was definitely a bit of venting that crept into my comments when I shared the news with my group. Fortunately, I did have the insight to lead with the why during that meeting, and the result was that our lab grew closer and more determined than ever as we navigated through a challenging situation together.

A more subtle type of vulnerability that can catch us by surprise is when we realize that someone we have mentored knows more than we do about a research topic, or is better than we are at a professional skill, especially when that knowledge or skill is something that *we* are supposed to be the expert in. While logic might inform us that this is inevitable—we can't possibly maintain the highest levels of knowledge and skill in every arena that our mentees will explore—it can still result in a feeling of vulnerability. In these moments, it is helpful to return to your "win" that you worked out in Chapter 3. At the end of your career, do you want to be remembered for always being the smartest and most skilled person in your research group, or do you want to be remembered for mentoring the next generation of researchers and for the impactful discoveries that resulted from your work together? There is a good chance that achieving your ultimate goal is going to require some vulnerability.

## ACTION ITEMS

- *Find time to share.* When choosing to be vulnerable, it is critical that you share what people most need to hear, not what you most want to say. Take a moment to think through the struggles or challenges that the members of your research group may be facing. Is it embarrassment at failed experiments, self-doubt about a future career aspiration, or a general lack of motivation? Now think of a story or experience in which you encountered and overcame a similar challenge. Depending on who needs to hear your story—one individual or your whole lab—choose the appropriate time to share and make a note to yourself to make it happen.

- *Assess your privilege.* Think back to the times when you felt vulnerable. Are there examples when your identities made the experience more painful or caused you to lose the respect of others because of something that you shared? Did you benefit from a privilege that made sharing safer or more comfortable for you? Now consider the individuals in your group and ask yourself how each of their identities will affect their ability to be vulnerable in their own current or future leadership roles. Remind yourself of this when you approach mentoring and career conversations.
- *Check your messaging.* As a leader, are you conveying to your group that leaders are born or that leaders are made? Can you think of times when you have told the group about your own growth as a leader? What evidence do they have that your skills and knowledge are things that you built up over time rather than something that just came naturally to you? Set aside a few minutes at an upcoming group meeting or social event and share part of your leadership journey, focusing on how you have grown over time. Find opportunities at least a few times per year to keep sharing parts of your journey with your group.

# 15

# CHALLENGING SITUATIONS

Unavoidable and Unpleasant,
but We Don't Have to Be Unprepared

"THERE'S NOTHING THAT I CAN DO ABOUT THAT—someone's behavior at work is totally separate from their job performance." I was stunned when an academic leader at one of my former institutions said those words in response to my concerns about inappropriate behavior that I and others were experiencing from a colleague. Though reeling in disbelief, I pulled myself together enough to respond with, "In that one sentence, you have just summarized almost everything that is wrong with academic culture." This moment also further cemented the realization that even when leaders can't fully shelter the members of their team from difficult situations, they have tremendous power to leverage such instances either for growth or for harm.

Now that I have the opportunity to serve in leadership myself as a department chair, I can definitively say that there is *always* something

CHALLENGING SITUATIONS

that can be done in that type of situation (and often something that *must* be done), and a person's behavior at work can *absolutely* be considered a part of their job performance. Moreover, these circumstances offer an opportunity to mentor others, so that they will be prepared to confront such challenges in the future when those who they lead are counting on them. Instead of telling me to ignore the behavior or develop better coping skills, the leader in my story could have provided support and guidance for how to deal productively with the situation and work toward a solution. This would have not only helped me cope with the challenge that was immediately before me but would have also provided a model I could use to support others in similar situations. Sadly, in that instance most of my learning was in the form of what *not* to do.

Perhaps you have already had a similar experience, or perhaps this story remains blissfully unrelatable. If the latter, it's important to recognize that even if you are fortunate enough to dodge such difficult situations in your career, it's unlikely that everyone in your research group will be so lucky. One needs to look no further than the statistics on the prevalence of harassment in academia and academic medicine to see that these challenges are endemic and that multiple identities, including race, gender, and sexuality, can increase one's risk of experiencing inappropriate behavior.[1] On a more hopeful note, the presence of leaders who actively promote a positive work climate results in lower reported incidences of harassment.[2]

If you are working to build a healthy culture in your research group, that is a great start for mitigating the negative experiences your students and postdocs will face. No culture is perfect, however, and your team members will have many interactions outside the research group, so there is still a reasonable chance that they will encounter a challenging situation that they'll need to manage. The situations they encounter may also extend beyond dealing with harassment or other inappropriate behavior. We've touched on many of these in earlier chapters—managing conflict, supporting someone in crisis, coping with poorly delivered feedback. In this chapter, we will move beyond talking about how to deal with these situations yourself and focus instead on how to coach others through the process. In doing so, you will support that individual as they manage the problem before them and also equip them to be a leader who can do the same thing for their future team members.

## Confidentiality Counts

As you support and mentor people through these challenges, it is natural to want to share with them all of the information that you have about a particular circumstance. This is, in principle, good leadership and mentorship—as we discussed in the previous chapter, transparency has been shown to increase both perceived trustworthiness and the effectiveness of leaders.[3] Moreover, navigating these situations is a problem-solving exercise and, similar to instances when you are working with a lab member to troubleshoot a failed experiment, the more information you can share, the easier it is to figure out what is happening and evaluate the options for moving forward. However, much as we saw in the "greater good" example in our discussion of ethics, there are competing interests that dictate how much information it is appropriate to share. You may also be limited by government laws or university policies that require you to keep certain information confidential. While maintaining confidentiality is critical at all times, it is especially important to recognize and respect these boundaries when working through a challenging interpersonal situation.

## CHALLENGING SITUATIONS

One of the more clear-cut—yet still difficult—circumstances is when there has been a complaint about inappropriate behavior, such as bullying or harassment. If the person who is acting inappropriately is someone in your research group, you are probably going to be directly involved in any corrective action that is taken, and at the very least, you are likely to know what action is being taken by your department or institution. If the person who is experiencing the behavior or who filed the complaint is also a member of your team, they are going to want to know what is being done to ensure that the behavior stops. This is where things get complicated. As much as you might want to share such information to help them feel reassured, in most institutional contexts this is considered confidential. But just because you can't share the specific consequences or action plan, that doesn't mean you have to be completely silent on the topic. A good approach is to let the person who is experiencing or witnessing the behavior know that while you can't talk about details, that doesn't mean nothing is happening. You can convey that they should expect to see the inappropriate behavior stop, and they should let you know if the issues persist.

There are other commonly encountered situations in which the boundaries are more variable and may be less clear. An example would be when one group member makes comments directed at another member during your lab meeting that could be considered uncollegial or overly critical. If this is the first instance, it may not fall into the category of bullying, but it is still something you should address so that it doesn't happen again.[4] When you meet with the offending individual and deploy your *Crucial Conversations* skills from Chapter 10 to figure out what is really going on, they may tell you they have been dealing with a personal emergency and are more on edge than usual.[5] Imagine you are now holding a parallel meeting with the group member who was on the receiving end of the uncollegial behavior. How much of the first conversation do you share with them? This is where the answer can vary. If the personal emergency involves protected information, such as needing to take a leave of absence for health reasons, that needs to be maintained as confidential. But if the information is something along the lines of the offending lab member feeling stressed out about their upcoming job interviews or frustrated by a recent manuscript rejection, this raises a tricky ethical question. One guiding principle is to consider whether you would be comfortable with that lab member knowing that you shared all or part of their story. This is both an intellectual exercise and a practical one, because it's very possible that they *would* find out. Also, if what they told you was overly personal, the individual you share this with might start to wonder what you say about them behind closed doors. Although there is no textbook answer for how much to say, one option that allows you to offer some context while still maintaining confidentiality would be, "This doesn't make the behavior okay, but this individual did express that they are dealing with a difficult personal situation, and they feel badly about how they reacted." Even better, ask the offending individual how much they're okay with you sharing. It's also important to state that the situations we're talking about here are instances when someone wasn't at their best, or an interpersonal interaction had gone awry. There is no amount or type of difficult circumstances that justifies behaviors such as bullying and harassment.

These examples focus on what you can do as a leader to navigate such situations, but they have two important implications for coaching others to lead through such challenges themselves. First, instead of being the one mediating the conflict, you can coach your group members through mediating the conflict themselves. Then you will no longer be the purveyor of confidential information, and the two people directly involved in the situation can choose what they feel comfortable disclosing to each other as they engage in dialogue to resolve the incident. Second, as you coach your lab members through the conflict resolution process, you have the opportunity to share with them the legal and ethical boundaries of confidentiality. Even if they end up in a workplace with different laws or norms, this can provide a solid foundation for understanding what they can and cannot share when they are leaders working through a difficult situation with their own team.

## Same Goal, Different Route

As you shift your focus from being the person handling the difficult situation to the one coaching someone else through the process, the methods and skills you will apply will be the same ones we've already learned. If feedback is needed, the Situation-Behavior-Impact (SBI) model is a valuable framework for crafting that feedback. If you are dealing with a conflict, then your first step will be figuring out what is really going on and what each person involved really wants. But instead of applying these skills directly yourself, you will be coaching your group member through the process, as they lead the way. They can think of this as the difference between going to a restaurant and having a chef plan and cook a meal for them versus taking a cooking class where they are walked through each step of the meal planning and preparation process. Both experiences end with a meal that's ready to enjoy, but one option teaches them how to go on and repeat the process for themselves.

In principle, you can teach someone a skill in real time as they are applying their new knowledge. If you've ever taught someone a new lab technique, you may have done this. On the other hand, if you have had this experience, you also know that the time spent modeling a technique is way more effective when the person learning it already has some background knowledge on the subject. Looking back to the polymerase chain reaction (PCR) example discussed in Chapter 13, talking someone through every step of what you are doing will certainly lead to better learning than if you just asked them to watch as you silently execute the process. However, their learning will be even greater if, before you ever set foot in the lab, you've given them an opportunity to read an article or watch a video about how PCR works. Similarly, coaching someone to lead through a difficult situation will be much more effective if they have already had the chance to learn the basic skills involved. This is where giving professional development talks in your group meeting or reading and discussing a book together can be especially helpful, and why, if you still haven't yet adopted this practice, now is a good time to start. Moreover, compared with other things you might teach your group members, the skills needed to resolve difficult situations are especially important to have in advance of when they are really needed. When they *are* suddenly needed, emotions and consequences will both be running high, making learning in the moment even more difficult.

## Four Coaching Contexts

Let's imagine that you have had the chance to teach your lab members some basic skills and knowledge for navigating difficult situations. We're now going to take a look at some specific examples we've discussed previously in this book, but instead of focusing on how you would lead through each situation, we will analyze how you could use the situation to help someone else develop as a leader. Most of the challenging interpersonal encounters that arise in the workplace can be grouped into one of four categories—conflict, inappropriate behavior, performance feedback, and crisis. We will dive into one example from each category and explore what it looks like to mentor someone as they lead through the situation. You may notice that managing these challenges requires multiple dimensions of leadership, from self-leadership to leading with peers or mentees to leading up with those in positions of authority. This is not atypical in difficult situations, and thus mentoring someone through such a circumstance also provides an opportunity for you to help them grow in each of these types of leadership and see how self-leadership, leading others, and leading up can be integrated to work toward a positive outcome.

*Conflict.* A student in your group is coauthoring a research paper with a postdoc, and although they thought the agreement was for them to appear as co–first authors, the postdoc sends the student a draft of the manuscript with the postdoc listed as sole first author and the student as second author. The student wants to have a conversation with the postdoc to see if they can resolve the situation before it becomes a conflict. In this instance, you can draw from the questions that we lined up in Chapter 10. First, ask the student what they want to get out of the conversation. This might seem obvious—they likely want to be listed as co–first author when the manuscript is submitted. Nevertheless, it is important to help them think through the relationship they have with the postdoc—are they close friends or two people who work together but have little more at stake than the casual "hello" they say in the hallway? Next, brainstorm together about what might actually be happening. Is this a simple misunderstanding? Did the postdoc agree to the authorship arrangement but forget to annotate the shared first authorship, or maybe they thought it wasn't important to worry about the author list yet? Or is the postdoc intentionally relegating the student to second author? If the latter, why might this be? Is it because they think the student didn't do their share of the work, or could it be misplaced jealousy because the student just won a major research fellowship for which the postdoc had previously been rejected? Just as when you troubleshoot research challenges, when you think through possible scenarios with the student, focus on those that seem most likely and make a plan for how they might figure out which is the real issue and what they will do next if this is the case. Finally, even though the student still doesn't really know what is going on, they can establish a mutual purpose to bring to the conversation with the postdoc. Ask questions like "What would it look like if you got everything you want out of this conversation?" You might help the student dig up some deeper answers beyond the authorship list—for example, if they want to be included more in the writing process or to have the opportunity for more transparent conversations about authorship in the future.

*Inappropriate behavior.* A student in your group who identifies as nonbinary is attending the department's summer picnic, and as people start to choose teams for the annual softball game, another student makes a comment about being unsure how to "count" the nonbinary student with respect to gender balance on the teams. If you are present when this happens, you can be an ally in the moment by either directly countering the comment or diverting the conversation to refocus on the softball game. You may also have a required role as a leader. In the United States, Title IX protects students from discrimination based on sex, gender identity, or sexual orientation, and as we discussed in Chapter 11, you may be a mandatory reporter and thus compelled to act in such a situation. In line with the theme of this chapter, you can also be a mentor to the student. Let's imagine that you are not within earshot of the comment during the picnic and the student later tells you about the incident. In addition to any required reporting, the student would like guidance for navigating the situation. In this case, your job is not to tell them what they should do but to help them understand and assess the options. You can start by validating their experience—just hearing the words "That wasn't okay and I'm sorry it happened" is a supportive way to start the conversation. You can then ask questions to help them think through what they want in the situation. Was this an isolated incident and they just want to make sure they aren't in proximity to the offending student again? Or was it part of a pattern of behavior that they want the department to deal with? Once you know the desired outcome, you can coach them through an action plan. Who do they want to talk with about the situation—the department chair, the campus ombudsperson, the Title IX coordinator, a member of the campus LGBTQ+ advocacy group, or another individual who can provide support and advice? Is there anyone they want to bring with them to these conversations as an advocate or for emotional support? Finally, you can help them decide how much time and energy they want to spend on this particular situation.

*Performance feedback.* A postdoc in your group is mentoring an early-career student who is participating in the postdoc's research project. The student starts missing scheduled meetings and is not completing the experiments they had agreed to perform. While you could step in and address this situation directly, you know the postdoc is planning a future career as a team leader and thus will undoubtedly encounter this type of challenge again. You have the opportunity to coach them through an action plan that they would then be prepared to deploy on their own in the future. As outlined in Chapter 9, you can help the postdoc get to a place of caring for the student and guide them in planning a few questions to explore whether the student has a life situation that might be causing their poor performance. Next, you can review the SBI model and help the postdoc write out their feedback for the student using that format. You can also discuss how to set up a meeting with the student. Does the postdoc want to share their feedback during a regular one-on-one meeting, or do they want to plan a special meeting with their mentee? Do they want to share a preview of their concern and feedback by email before the meeting so that the student can prepare for the conversation, or has the situation reached a point where it is better to wait for the meeting? Once you've coached the postdoc through their plan for the meeting, you can also review how they have set and communicated expectations for attending meetings and performing experiments, and explore whether there are any areas where they might want to increase clarity. This may or may not prevent a similar situation from happening in the future, but either way, the more clearly that expectations are communicated, the easier it is to have a conversation about them when they are not being met.

*Crisis.* A graduate student in your group is serving as a teaching assistant for an introductory class, and an undergraduate student enrolled in the class visits their office hours and shares that they are struggling with anxiety. As you talk with your group member about this, start by helping them assess the severity of the situation. If the undergraduate said that they are just feeling a bit stressed out about the upcoming exam, you can skip ahead to the coaching conversation that follows. If, however, the graduate student gets any hint that the undergraduate is at risk or might be considering self-harm, the next and most important step is to guide your group member through the reporting process. In this case, they should first report the incident to the course instructor. If that individual does not seem responsive, they can also reach out to the department's director of undergraduate studies or the department chair. In parallel with any reporting, if your group member wants to follow up with the undergraduate student to offer support, you can help them plan that conversation. Do they want to do that by email or by inviting the student to stop by again during office hours?

You can also work together to research the mental health resources that are available on your campus—no matter how mild or severe the particular situation, providing the student with information about these resources is always a good idea. Perhaps most important, you can help your group member set boundaries, both with respect to how much time they are willing to give to the situation and for differentiating between the support they can provide as an ally and the counseling that should be done by a mental health professional.

## Where Problems Go to Die

One of my friends who is an academic leader was talking about hiring a new manager for their department and said that they had told the person who got the job, "I want your doorway to be the place where problems go to die." The complex organizational structure in academia can make it especially easy for people to say that a problem is not theirs to deal with and to keep passing it on to someone else. My friend had recognized the inefficiency and frustration that this causes, and that overcoming it requires that there is a person who ultimately takes responsibility for getting problems solved. In the case of your research group, that person is you.

I hope you have seen the value in coaching someone else to lead through challenging circumstances, but just because you've empowered them to take the lead doesn't mean that you no longer have a responsibility to see the situation resolved. As you work together with your group member to formulate a plan, you can also decide when and how you will follow up, to ensure that they continue to feel supported until their challenge is resolved. If the situation is minor, such as the TA whose student reported feeling nervous about the upcoming exam, your follow-up plan may be to let your group member know to contact you in two weeks if the situation doesn't seem to be getting better. Or the follow-up may be a specific part of the plan, for example if you set an appointment to debrief after your group member has had a conflict-resolution discussion with a colleague or met with their mentee to discuss time-management habits. For complex situations that may take the longest to resolve, you can plan regular check-ins, perhaps by offering to make your monthly one-on-one meetings a half hour longer so you can discuss the progress on the situation and work together on a plan for the future. These follow-ups will not only show your commitment to seeing the challenge resolved but also provide continued opportunities for mentoring those you lead, so that they can continue to grow and learn how to navigate challenges as a leader in the future.

## ACTION ITEMS

- *Create confidence with confidentiality.* You may have had to take required training for handling confidential information, such as student grades, but there are many types of information for which your university policy may be less clear. Set a meeting with a senior colleague, a member of your department's leadership, or a university representative such as the ombudsperson or a member of human resources, and ask them about situations you've encountered or can imagine encountering (such as a lab member taking medical leave) and what information you can and can't share in such contexts. Ask if there is a university website on which you can find these policies and perhaps a flowchart or other resource explaining what information you're required to report to a university official and to whom you should report it. Take notes and keep them in a convenient place so that you can refer back to this conversation when one of these circumstances arises. For situations when there is no university policy, think about your own ethical guidelines and decide in advance what you think is appropriate to share and what information you will hold in confidence.

- *Act it out.* Coaching someone to lead through an interpersonal challenge can equip them with a plan, and possibly even a script for the next conversation that they need to have. Still, it is a big leap to go from knowing the words to say and delivering them in the best way in a conversation. Next time you find yourself coaching someone on how to navigate a difficult situation, ask if they would like to try rehearsing the conversation. This role-playing may feel a bit awkward (and you can make it less so by acknowledging that), but it can be worth it if it helps them get it right in the real conversation, when the stakes are highest. To make the exercise feel a little less intimidating, set a timer for five minutes, or even just three, so you both know you only have to keep the conversation going for a short time.

- *Follow up.* Think about a challenging incident that one of your group members has brought to your attention in the recent past—this could be a work situation or a personal emergency, such as a sibling getting into a car accident. Make a plan to follow up with them to ask how things are going and if you can do anything to support them in navigating the situation. Depending on the sensitivity of the discussion, plan to ask them the next time you see them, or send an email inviting them to meet over coffee in your office or some other semiprivate location.

# 16

## EMPOWERING

## The Gift of Opportunity

MY ELDEST CHILD HAS JUST RECEIVED his learner's permit and is about to start driving. As we think about how he will go from learning where the ignition is and how to operate the windshield wipers to being able to drive himself to college in a few years, one thing is clear—no matter how much we talk him through the process, learning will require getting behind the wheel and trying it for himself. The same thing applies to leadership.

Your introduction to leadership may have felt like the opposite experience—someone threw the keys at you one day, told you where the car was parked, and then expected you to drive to a faraway destination called "tenure" and arrive there on time, all with minimal instruction or behind-the-wheel training.

As we have seen throughout this book, developing strong leadership skills involves a balance of both learning and doing. Just as your experience as a new faculty member was not ideal because you were asked to lead before you were really taught how, it is not ideal to only teach your group members but not give them a chance to test out their skills in real leadership situations. As we talk about how to help those we lead develop as leaders themselves, it's important that we not only provide professional development content to teach them key skills but also give them a chance to serve as a leader so they can practice those skills and see how they translate to real-world situations.

Our role is to figure out how much guidance each person needs as they start to practice their own leadership. To continue with the driving analogy, when I was first learning to drive, I had someone sitting in the seat right next to me giving me instructions at every turn and ready to help out if I made a mistake. This was necessary for my safety and that of others because I was changing lanes and merging with other moving vehicles for the first time. When I later learned to drive a car with a manual transmission, my dad gave me the keys to his very old truck and sent me out on my own in a deserted parking lot. My face must have betrayed my unspoken question—"You're not coming with me?"—because he said, "There's little damage you can do to that truck at this point and there's nothing around here that you can hit—just go for it." As we'll see in this chapter, some leadership opportunities have higher stakes than others, and how we approach mentoring someone to lead can be adjusted accordingly.

In the previous chapter, we started to see what it looks like to empower the individuals in our research group to be leaders themselves. Whereas those examples focused on specific situations that arise unexpectedly, in this chapter we will talk about how to intentionally build opportunities for people to take on leadership roles, and how to find the right balance between providing guidance and allowing them to learn from their own experiences. This is your chance to provide your group members with the behind-the-wheel leadership training that will prepare them to hit the highway and merge into the fast lane with confidence when the time comes for them to move on to their own independent careers.

## Not Everybody Is You

What may seem obvious but can be exceptionally difficult in practice is to remember that the people you are mentoring are not you, and the goal is not for them to become a replica of you. Coaching others to develop as leaders requires that we recognize their unique identities, personality traits, and career goals and then help them to be the best version of themselves in their leadership journey.

As you consider the opportunities that your group members might want to explore for practicing leadership skills, it's essential to recognize that each individual's identities can dramatically affect how they experience leadership and the style with which they will choose to lead.[1] As we discussed in Chapter 14, for example, individuals who are not aligned with the majority identity in their position are more likely to have their competence questioned. This can impact the leadership situations in which they feel most comfortable, and it should also influence the level of support and validation you give each person as they take on a leadership role in your group. Interestingly, for those who are growing as leaders, their perspective on the effect of their identities can change over time. In a study of college students who were taking on leadership roles, researchers tracked female students' views about women in leadership as a function of their own leadership development stage. They found that the students were aware of biases about women in leadership from an early stage in their own leadership development, but they did not internalize these biases as relating to their own leadership journey until the later stages of their development.[2] Members of your research group may experience a similar evolution in their view of how their identities intersect with their leadership. Even if you don't share their identities, you can serve as a listening ear and a supportive thought partner as they work through this process. You can also help to connect your group members with other leaders who do share their identities and thus may be able to serve as role models or provide advice from their own personal experiences.

In addition to different identities, your lab members also have different personality types and career goals that are important to consider. Thinking back to the personality assessments we discussed in Chapter 2, recognize that just as you have specific strengths and preferences, so do those who you lead and mentor. If you are an extrovert who loves to be in highly visible leadership roles, you may be mentoring individuals who share this preference, but you may also mentor individuals who are more introverted and prefer to lead from behind the scenes whenever possible. Or you may have a strength in thinking about the big picture and creating the initial vision for a project, whereas the individual you are mentoring finds their greatest strength in planning out the details to implement and troubleshoot that project.

Research has shown that in many dimensions, the more identities and personality traits we share with an individual, the more effective we can be as their mentor.[3] This is critical to note, because your goal is to provide equitable opportunities for each person in your group. This means that while the easiest and most fulfilling mentoring discussions you have may be with those who you have the most in common with, you will probably need to spend more time (and energy) with those who are different from you in order to give them the same level of mentorship. It is also important to recognize that providing equitable opportunities for leadership doesn't mean giving each individual the same opportunities. Rather, you can tailor the leadership roles and tasks that you offer in order to best suit each person's strengths and career goals, and in doing so, you will be providing each person with an equal opportunity to develop themselves as a leader.

## Sharing the Driver's Seat

Leadership development is a complex topic, but taking practical steps to allow those in your research group to grow as leaders doesn't have to be overly complicated. Returning to our driver-training analogy, I could give my son some practice by having him drive to arbitrary places around town, but it would be even better and more expedient if I let him take the driver's seat for trips that are already part of our plans, such as going to the grocery store or his football practice. Similarly, when it comes to empowering leaders in your lab, rather than trying to create an entirely separate leadership development program, you can leverage the tasks and responsibilities that are already part of your plans.

As we've discussed, there are certainly some aspects of the faculty job that require confidentiality or that you are legally or ethically obligated to perform yourself. Yet a significant portion of your day is likely filled with tasks that offer opportunities for developing others as leaders. Empowering your team members to take over responsibility for these efforts not only benefits them but can also produce better results. Moreover, the time you save by not having to do the work yourself can be put back into mentoring those who have taken on leadership of these tasks, thus multiplying the impact of your time with that mentee. The specific tasks that you can delegate will depend on your institution, your research discipline, and your career stage, so I will outline several options. As you read through them, think about those that are most relevant to your own position and research group, and how you might give someone else in your lab the opportunity to try "driving" that task, rather than doing it yourself.

*Mentoring.* The most obvious leadership opportunity you can give members of your group is that of mentoring others in the group. If your group is larger than two or three people, then it is likely that this is already happening in some form, and you can formalize the process to equip and empower those who are serving as the mentors. Mentorship within your lab can involve a variety of tasks and commitment levels, ranging from reading and offering feedback on a manuscript draft to serving as the primary point of contact for a project over several months or years. While this mentoring structure may already be in place informally, your lab members could still benefit from greater clarity and transparency about the roles that more senior members can step into as well as where newer members can go for feedback and the types of advice they can ask for. Take time during a group meeting to discuss the forms of mentoring that already exist and what would be helpful to add to your lab practices. Involving the group also creates a mechanism for you to leverage your time to mentor your lab members on their own mentorship. For example, if a postdoc is mentoring an undergraduate student working on their project but the research doesn't seem to be getting done on time, you can work with the postdoc to create a system for setting goals and creating accountability with their mentee. This might consist of having the undergraduate student email the postdoc a set of weekly goals each Monday and then a slide deck of results each Friday. Whatever the situation and the system implemented, the important point is that you are coaching a member of your group to be a leader.

*Hiring.* If you have a group, then you have some type of hiring or recruiting process. A good first step in empowering members of your group is to involve them in the decision-making about how you evaluate candidates and the choice of which ones to recruit. Set aside a meeting time with your group to discuss the skills and strengths you are looking for as you evaluate candidates. If you've outlined your group values, then this becomes a much more straightforward exercise of simply adding in the technical capabilities that you are hoping the new hire will bring to the group. Ensure that group members have an opportunity to meet and interact with prospective team members, and then schedule a meeting to discuss the candidates as a group. You can start the meeting by reminding everyone of the specific strengths you are looking for and reiterating the shared culture of the group, but it's important to hold back your own opinions of the candidates until the end. As the leader, your words carry extra weight, and identifying your top candidate can give the impression that your decision is already made. Instead, keep the discussion focused on the specific strengths of each candidate, watch out for and mitigate implicit biases that might work their way into the comments, and ask thoughtful follow-up questions that help to surface the information that will be needed to make the best decision. Over time, you can consider letting someone in your group lead this process. You might have them put together the interview schedule, choose questions to ask during meetings, and even facilitate the group debrief meeting. Your role then becomes that of a backup—you are there when needed but giving someone else the opportunity to experience leading in the recruitment process.

*Equipment.* Depending on your area of research, you may have a lab with equipment. If you do, then you definitely have tasks that can be turned into leadership opportunities. When I first started my lab, I was the only member of the group, so I made all of the decisions about what to purchase. When we moved to a different institution seven years later, I turned the leadership of this process over to the group. I outlined what items I thought we needed to buy and an approximate budget for each. Then, group members formed teams for each item, researched the options, solicited quotes from vendors, and came back to the group with a recommendation for what we should purchase. In some cases, the group discussion moved the decision to a piece of equipment beyond what the team recommended, serving as a lesson that the leader's job is to help the group find the best option, even when it is not their initial choice. Importantly, the teams said that this exercise gave them valuable experience for their future roles, even if they don't stay in academia. You may not have an opportunity or a need to reequip an entire lab, but if you do have equipment requirements, you can implement a scaled-back version of this process when it comes to purchasing service contracts or new accessories. You can also appoint group members as leaders for managing equipment maintenance and training for new lab members and empower them to create systems such as standard operating protocols, a logbook, or a scheduling calendar to best manage the instruments.

*DEIJ.* Creating a diverse, equitable, inclusive, and just environment in your lab is your responsibility, but that doesn't mean that you are the only one who can (and should) do the work to achieve this goal. Our lab holds regular discussions on topics related to DEIJ, and the responsibility for choosing the topic and leading the discussion often rotates among lab members. We also revisit our lab policy document every year during our lab retreat. One lab member volunteers to lead this session, during which they facilitate a discussion on what policies we may need to add, edit, or remove to better align the document with our desired lab culture. Group members have the chance to lead in the crafting and documenting of these policy changes in our lab manual. Depending on the size of your group, it may also be helpful to appoint a diversity officer or to form a DEIJ committee that can serve as a conduit for feedback on the environment and culture in your group and take the lead in proposing new policies or practices to improve equity and inclusion.

*Teaching.* If classroom teaching is part of your job, there's a good chance that at least some of your lab members have also participated in your department's teaching mission by serving as a teaching assistant or in a similar role. Teaching opportunities are not limited to the classroom, though. If someone in your lab is an expert in a specific technique that others want to learn, you can empower them to design and lead a short "boot camp" to teach the skill to others. In our lab, this has resulted in members running a "cell culture camp" and a "microscopy camp." You can also include your lab members in providing professional development content for your group—this can involve consulting with them on topics, getting their advice on content, or enabling them to lead a session at a group meeting on a topic of their choosing.

*Finances.* No matter the size of your research group and your discipline, you have to manage a budget to pay for people and stuff. Whether this means planning for annual travel for you and one student to visit an archive, or being responsible for a team of twenty grad students and postdocs and all of the materials and supplies they need for biomedical lab work, you can involve your group members in the process. As you review your annual budget and make projections, include them in the meeting. Ask them to take a look at the previous year's expenses and determine if the group is coming close to what is budgeted and if they notice places where changing vendors or the ordering strategy could result in savings. Importantly, include them in the discussion of how much new funding is needed in the coming year to sustain the lab's planned activities, and where you might seek out that funding.

*Grant writing.* In the likely event that your budget projections indicate that you need to apply for new funding in the next year, this is another opportunity for your group members to lead! One of the most valuable professional development exercises our lab ever embarked on was creating "reverse outlines" of each of the major types of grant proposals we write. Working in groups during one of our annual retreats, we picked apart examples of successful grants in order to identify the purpose of every section, paragraph, and sentence. The result was a set of templates that outline in detail what content needs to go where in a proposal. With these templates in hand, it became possible for group members to effectively participate in and even lead the drafting of new proposals, which in turn has informed the structure for many of our recent group retreats. To provide support and scaffolding for this exercise, we meet as a group to discuss the science that will go in the proposal and to empower individuals to assign themselves sections to draft. We then provide feedback throughout the writing process to both sharpen the proposal and help each individual develop their grant writing skills.

## Learning in Levels

Just as I wouldn't take my fifteen-year-old out on a busy highway or on a week-long road trip for his first driving lesson, you don't want to launch someone into a high-stakes situation for their first leadership training. Instead, think about low-stakes (and ideally low-commitment) activities that can serve as good entry points to developing someone's leadership skills and identifying their strengths and interests, and then build from there.

As an example, let's turn back to the topic of mentoring. It is not uncommon in the lab sciences (and this has happened way too often in my own lab) that a grad student's or postdoc's first real experience mentoring another researcher comes about when they agree to work with a new student in the group. In these cases, the senior member of the group becomes the primary lab mentor for the student, and the commitment can be a years-long undertaking—so, both high-stakes and high-commitment. Despite the lack of a good leadership on-ramp, these situations are often extremely positive and rewarding for both mentor and mentee. But there are also times when the senior member realizes that this type of mentoring just doesn't align with their strengths and interests, and in these cases the experience is far less constructive.

Avoiding this situation is straightforward, but it requires intentionally building in lower-stakes and lower-commitment opportunities for people to grow their leadership skills and explore their interests. And this can be a win-win, because creating additional opportunities for people to be a mentor also creates more ways for individuals in the group to receive mentoring. In our group, the leadership opportunities in mentoring allow for graduated levels of commitment:

- *Teach a technique.* As new students and postdocs join the lab, there are techniques that we use that they haven't learned yet. Experienced lab members can spend an afternoon teaching a new member how to run an instrument or perform a protocol.

- *Guide a rotator.* In our graduate program, students have a chance to "rotate" through labs for a few weeks as they consider which one they want to join. Our group has devised a system in which each rotator is assigned a rotation mentor—an existing member of the lab who can serve as their guide and go-to person during those few weeks.
- *Support a summer opportunity.* Those who enjoy mentoring others but are not sure they are ready for a long-term commitment can look for shorter-term opportunities. Perhaps your group is hosting a high school or undergraduate student only for the summer. This can be a perfect way to get the full experience of mentoring someone in the lab, but with the knowledge that it is only for a few months.
- *Be a long-term mentor.* This is for group members who have advanced through the earlier levels or have had other experiences that make them ready to commit to a long-term mentoring relationship. Mentoring a new lab member involves walking them through the different stages of learning, from figuring out where supplies are stored to crafting their first research presentation to making project decisions. The commitment can extend over multiple semesters or even years, but this greater time investment also translates into greater impact and thus an even more rewarding experience as a mentor.

These different mentoring levels provide a progression, but they don't need to be overly prescriptive. If someone has been a rotation mentor, clearly enjoyed that role, and is prepared for a much more significant commitment, then it could make sense for them to skip a level and dive into a long-term mentoring role. Additionally, there are probably many more levels and mentoring opportunities in your group than you could articulate in a list or flow chart. Perhaps most important, this approach can be applied to any type of leadership task that you are empowering someone to try. Since the goal is to help each individual in your lab move along a unique leadership development path, there's no reason why the opportunities they explore can't be similarly unique and aligned to their specific interests and career goals.

As you provide leadership opportunities for those in your group, you can take advantage of many of the systems and policies that are already in place to help your lab members be effective in these roles. For example, our group has subgroup meetings every three to four weeks and we have a specific slide template that outlines previous goals (and progress made toward meeting them), key data obtained since the last meeting, and goals for the next three to four weeks, until the subgroup meets again. Since the established members in my group are already very familiar with this format and it's documented in our lab policy manual, it is easy to adapt. When they start mentoring an undergraduate student, I often suggest that they use a modified version of the template for weekly updates, to help their mentee keep track of goals and progress and receive more timely feedback. Excitingly, this adaptation of systems and policies doesn't need to be a one-way street—given the freedom and encouragement, your group members will develop their own systems and policies as they take on leadership roles, and many of these can be integrated into the organization of your entire lab.

## Nobody Is Perfect

It is impossible to learn a new skill without making mistakes, and this *definitely* applies to leadership. The leadership development approach I've shared can help set people up for success by gradually increasing the scope and responsibilities associated with their leadership roles, but it is inevitable that at some point they will drop the ball on something that needed to get done, or manage a tough situation in a way that makes it worse. Great leaders aren't people who never make mistakes; they are people who can learn from them.[4] As someone who is working to develop others as leaders, your job is to be prepared for these instances, to resist the urge to "grab the wheel" and take over, and to instead provide candid feedback and helpful support so that your group member can learn from the experience and keep developing as a leader.

Mistakes can be particularly powerful moments for you to deploy vulnerability, as we discussed in Chapter 14. The members of your group can no doubt find many people who can relate to making a mistake in their research, but they will have far fewer role models who can relate to making mistakes in leadership. Whether they have misinterpreted the instructions for a grant application and left out an important component of the proposal or caused a conflict by accidentally delegating the same equipment maintenance task to two lab members, there's a good chance that you have made a similar mistake in the past. Sharing your story can normalize failures and help that individual see that making mistakes is a natural part of leadership, and in turn promote greater innovation and better long-term results.[5]

In parallel with normalizing mistakes, you can teach your group that good leaders create a plan to correct for their mistakes, work to fix the damage done, and learn how to avoid that same mistake in the future. In these moments, it is particularly important to lean into the coaching style of leadership. As you're talking with a group member who has made a mistake in leadership, the path forward may be clear to you, especially if you've also made that mistake before. It will be much more helpful, though, if rather than just telling them what they should do, you coach them through the process of figuring it out for themselves. Ask guiding questions, such as, "What do you think is needed to repair this situation?" and "What would you do differently next time?" In doing this, you not only help them to make things right in the situation they are managing in that moment, but you also give them the skills to repair and learn from their mistakes throughout their career as a leader, and that is what expert-level driving looks like.

## ACTION ITEMS

- *Reassign a task.* Look at your to-do list and choose a task that you could empower someone else to take ownership of and lead. If the person taking on the task has not yet had the opportunity to lead in that area, consider how you might break the task into levels, and offer them a low-stakes, low-commitment entry point. You can also consider what systems or practices you have created for your group that the individual could use as they lead in getting the task done.
- *Make it a policy.* Think about a task that you have already successfully transformed into a leadership opportunity for those in your lab—perhaps it is mentoring summer research students, managing equipment, or organizing postdoc interviews. Take a moment to check whether this leadership role is outlined in your policy manual, and if it's not, include it! By writing these leadership opportunities into policy, you create equity by ensuring everyone in the group can see what opportunities are available, and you can also provide some initial instructions and scaffolding for those who take on each role.
- *Keep it moving.* Just as it's important to customize leadership opportunities to an individual's strengths and future ambitions, it's important to diversify their leadership experiences so that they gain a range of different skills. For roles that are limited to a

set number of people at a time (such as safety officer), decide what the term limit should be and create a rotation so that multiple people in your group have the opportunity to experience that role, and the person currently serving in it also has a chance to experience other areas of leadership over time. For roles that have more ad hoc participation (mentoring, retreat organization), check in with each of your group members the next time you have one-on-one meetings or review individual development plans, and discuss what skills they have gained and which they still want to develop while in your group. Make a plan for when and how you can create those opportunities.

# 17

# COMMITMENT

## Be in It for Life

"BEFORE YOU DO ANYTHING, why don't you talk with your PhD advisor?" This suggestion from my spouse, John, stopped me in my tracks and completely changed the trajectory of one of the worst days of my life. And when I did get a chance to talk with Jeff Moore—an expert in chemistry, an experienced leader, and also my graduate school thesis advisor—my outlook on the situation was transformed from despair and confusion to confidence that I had a plan and a support network. What often surprises people is that this conversation took place ten years after I finished graduate school.

It was the day of my department tenure vote, and the outcome was not what I had expected. As an assistant professor, I was acutely aware that there are no guarantees when it comes to tenure votes. At the same time, I had received strongly encouraging comments from a number of senior researchers in my field who would presumably be writing my external evaluation letters, and people would roll their eyes when I (somewhat) jokingly mused about the careers I would pursue if I failed to get tenure. Everything seemed to be lining up for me to clear this hurdle and move on to the next stage in my career . . . and then half of my colleagues voted to deny me tenure.

I was devastated. This vote would likely lead to the loss of my job and could mean the end of my academic career. And it felt so much worse that I hadn't seen it coming. As I absorbed the results of the vote, I had to contend with the thought that perhaps the warning signs that I wasn't actually on the right track toward tenure had been there, and I had somehow missed or ignored them. The worst part was knowing the disruption that this vote would cause for my family and my research group—I felt like I had failed the people who matter most to me and that my mistake was going to have a very real negative impact on their lives.

As I sat in the car crying and with words spilling out of my mouth about what I needed to do next, John offered the suggestion that I talk with my advisor. I went silent and then said, "That is *exactly* what I need to do." I sent Jeff an email to share what had happened, and almost immediately received a response with words of encouragement and an offer to talk whenever I was ready. When we did connect by phone, he offered some candid thoughts about what had transpired, asked me a few questions about how I wanted to move forward, helped me create an action plan for the coming days and weeks, and then outlined how he could offer practical support as I implemented the plan. It was one of the most effective examples of mentoring I've ever experienced, and it came in the moment when I needed it the most.

As my tenure process continued to unfold over the coming months, I received validation from various committees and academic leaders that the negative votes were indeed anomalous and that I had in fact met the requirements for tenure, and I was eventually granted the promotion that I had earned. While my tenure story still ranks among the worst experiences of my life, it is simultaneously the most important event in my entire career. As I moved past the hurt and chaos and emerged as a tenured, midcareer faculty member with a successful research program, I found myself feeling more bold and fearless than ever. I was also a more compassionate and empathetic leader, and had formed a vision and drive to create change and help repair the brokenness in academic culture. Much of this was possible only because I had great mentors like Jeff, who supported me throughout.

I also want to highlight that receiving support from mentors isn't limited to bad situations, and the mentors you can turn to aren't limited to those in whose research groups you have worked. In parallel with writing this book, I found myself interviewing and negotiating an offer for, and then accepting and moving into, a department chair position. While there are many blogs and advice columns about how to land your first faculty position, there is no published playbook for interviewing as a "senior hire." Moreover, the stakes were especially high because the department chair role meant I was negotiating for both myself and my department. I was again grateful for the opportunity to call on Jeff and other mentors in my life who could speak to me confidentially and help me work through each step in the process. And now that I'm in the role, I continue to lean on these mentors. When I encounter a situation that I'm not sure how to handle, I look to my "phone a friend" list of current or former department chairs and reach out to one or more of them for advice.

No matter what your current leadership positions include, you can probably relate to the increasing complexity of the situations you need to manage, and thus the increasing need for mentors as you advance in your career. The same will be true for your group members. The frequency and immediacy of your mentoring will decrease as they graduate and move on from your research group to their future careers. But the importance of your mentoring can increase as your group alumni also find themselves confronting increasingly complex situations. This underscores the privilege we have as research advisors not only to be leaders and mentors for our lab members now, but also to be there for them throughout their careers. This may sound like it takes some time, but the reality is that it can return even more in happiness and fulfillment.

## Mind the Path

When it comes to coaching or developing others, a critical mistake that leaders can make is trying to serve as a role model when what someone really needs is a mentor. Although these terms tend to be used interchangeably (and it's possible to be both a role model and a mentor), they do have very different meanings. A role model is someone who another individual aspires to emulate, for example in their career path, skills, or achievements. In contrast, a mentor is someone who helps an individual figure out what career path they want to take, what skills they want to develop and how to do that, and what they ultimately want to achieve. In rare instances, one of your lab members will say that they want to be exactly where you are or to attain the same skill or level of achievement that you have. In these cases, you can serve as a powerful role model and share information about your path and how they might follow similar steps for themselves. In all other cases, your job is to recognize that their goals are different from yours and to serve as a mentor to help them figure out what *they* aspire to and how to best pursue those goals.

In practical terms, the difference between being a role model and a mentor shows up in the words that we use to frame our advice. Consider a situation where a graduate student in your lab has been interviewing for jobs and they have two offers. One is at a large company with good pay, but it's located in an expensive city that is far from their family. The other offer is at a smaller and less prestigious company and it pays less, but it is an hour's drive from their family and has access to all of the outdoor activities that align with their hobbies. If you were acting as a role model, your advice might sound like, "When I was in a similar situation . . ." or "What I would do is . . . ," whereas being a mentor means asking questions, like "What do you view as the main positives and negatives in each choice?" or "Where do you want to be in five years and how would each option help you get there?" Again, the key is recognizing that in the majority of situations, your lab members are looking for a mentor rather than a role model. In these instances, your job is to help them make the best decision for themselves and to chart their own career path, not to replicate yours.

Recognizing the unique goals and aspirations of each of your group members is also relevant to how you measure your own success as a leader. Academia has the strong tendency to define success by the number of people we mentor who then go on to careers in academia. While this attitude varies across disciplines and institutional contexts, it is universally unhelpful. It is great if someone you lead wants to pursue an academic career, and it should be just as great if they want a career in industry, government, nonprofit work, or anywhere else. If you are reading this book, you are someone who cares about the people you lead, and you probably already want your success to be defined by the number of people who are able to pursue the career of their choice, rather than by the number who follow directly in your footsteps. Still, the message we receive from academic culture that some career paths are more valuable than others is so loud and constant that it takes equally frequent and forceful reminders to keep our true definition of success in clear focus.

## Checking In

Once you've focused your definition of success as a leader on helping each person in your lab to pursue their career goals, one of the best parts of this job is watching them do exactly that. Like many of the concepts we've discussed in this book, keeping in touch with those you have mentored doesn't have to take a lot of time, but it does take intentionality.

You can lay the foundation for this with the messages you communicate to group members while they are still working with you. If you do have alumni who have moved on from your group and you have the chance to see one of them at a conference or they call to catch up, mention it to your current group and talk about how much you enjoy these opportunities. If you don't yet have lab alumni, you can talk about how much you enjoy keeping in touch with your former mentors and how you hope that your current group members will do the same when they move on to the next stage of their careers.

You can also be intentional in seeking out opportunities to connect. When you travel for conferences or seminars, or to sit on review panels, think about whether any alumni from your group are living near the location you will be visiting and make a point to reach out and ask if they would like to meet up. If you are able to connect, you'll experience the fun of hearing about what is happening in their life and career, and even if you aren't able to coordinate schedules, you'll have shown that you care, and it may open the door for a brief email update or future video chat. My friend Lou Charkoudian, a chemistry professor and an expert mentor, has a brilliant approach for involving her entire group in staying connected. Charkoudian organizes a virtual lab reunion every year that not only allows her to connect with lab alumni but also allows them to catch up with each other and with current members of the group. She says this practice is especially powerful for faculty who, like her, are at a primarily undergraduate institution, because her lab alumni pursue a highly diverse array of career paths after graduation.

Finally, you can initiate a conversation when you know that a former group member is at a critical career stage. When I was nearing the end of my first year as a faculty member, Jeff Moore reached out and said, "We should catch up." In that phone call, we covered a number of topics, ranging from how I was doing with recruiting students to when I would publish my first paper to how to network with leaders in my field for award nominations and tenure letters. Near the end of the call, he underscored that I could reach out to him whenever I ran into a situation for which I needed advice but didn't want to go to someone in my department. At the time, I thought, "I can't imagine what type of situation that would be . . ." Needless to say, I have taken him up on this offer many times since that initial conversation.

There are many memorable aspects of that first catch-up call with my PhD advisor, but the one that stands out to me the most is what he said at the end of our conversation. It went something like this:

JEFF: You know you can reach out to me anytime you need to talk.

ME: I know, and I really appreciate that.

JEFF: I mean it. I know you think I'm busy. I am busy. But this is important, and I'll make the time.

ME: I know. I'll do that.

JEFF: I really mean it. No matter how busy I am, I can always make time to talk.

This conversation is especially important to me now, as I am the one in the advisor role. What Jeff had recognized and powerfully broken down is one of the biggest barriers to keeping in touch—perceived busyness. As you try to stay connected with your own alumni, they may be similarly hesitant to reach out because they know you are busy. In whatever words you want to use, be intentional in expressing the same sentiment that my advisor did, to assure them that no matter how busy you are, they are a priority and that you want to make time to talk.

## Evolving Roles

One of the other great joys of keeping in touch with your former group members is seeing how your relationship evolves over time. When they move on from your group, your evaluative role almost completely evaporates and you become far less involved in the day-to-day of offering advice and guiding them in their research. This allows a new type of relationship to emerge—one in which you are still an important mentor, but you can also be a peer. You may find yourself at a conference, watching a former group member present on the research they are now doing as a leader in industry. Or for the alumni who stay in academia, you may have a chance to visit them at their new institution and meet their current group members—your "academic grandchildren." These experiences number among the most fun and heartwarming moments in my own academic career, and even though conversations generally grow less frequent when people move on from the group, the peer relationship enables these interactions to take on even more depth and meaning. It is also very possible that as your former group members move into their own independent careers, they will develop skills that are different from yours, and you may even find yourself benefiting from *their* mentorship.

## Be a Lifelong Learner

In parallel with the fulfillment to be found in being a mentor for life, the positive impact you can have on others will be maximized if you approach mentoring and leadership with the mindset of a lifelong learner. As we've acknowledged throughout the book, leadership involves people, and people are complicated, unique, and always changing. Thus, leadership skills are something that we can continually improve over time but will never truly master or perfect, and that is part of the fun—even at the later stages of your career, there is still space and potential for growth.

As you look to the future, take a moment to appreciate how far you have come. When you first set your sights on an academic job, you may not have expected that to mean a leadership job. You may have found yourself, like me, in a job that you weren't prepared for. Think back to the knowledge and confidence you had at that point, or even when you started reading this book. Now reflect on what you have learned since then, what leadership skills or situations you feel more confident about, and how your growth has had a positive impact on those you lead. As you continue to learn and grow—whether by seeking out additional books, podcasts, and other resources, gathering feedback from those around you, or maybe occasionally rereading portions of this book—you will continue to improve the experience of those in your research group and help each person perform at their best.

And your impact doesn't stop there. The individuals you are leading right now will become the next generation of leaders, and what they are learning from you now is creating a model that they will take with them into their own future leadership roles. As you learn to be a better leader and how to better mentor future leaders, your actions have a ripple effect that will go on to affect more people than you can probably imagine. We're in this together, and together we can change academic culture and create a healthier environment for generations to come.

## ACTION ITEMS

- *Reach out.* Make a plan to get in touch with a former mentee so that you can hear about how they are doing and offer continued support. Look at your upcoming travel schedule to see if you will be visiting a location where any of your former group members are now living. If you will, send a message to see if they have time to meet up for a meal or coffee. Alternatively, think about whether any of your former group members might be at an important career juncture, such as finishing their first year in a new role, and contact them to ask whether they want to have a quick check-in conversation by phone or video chat. For bonus points, schedule a virtual group reunion to catch up with alumni and connect them to your current lab members.
- *Circle back.* Just as you would love to get updates and good news from the people in whom you have invested your time, your mentors probably feel the same way. Get in touch with a

former mentor via email or text and let them know about something exciting that is happening in your career or an area where you could use advice. If you had a positive relationship with that mentor, this can be a great opportunity to use your Situation-Behavior-Impact feedback skills to let them know that something they did or taught you is now having a positive impact on your career. There is a good chance that your message will make their day.

- *Pass it on!* Now that you've finished reading this book and have had a chance to work through some of the action items and test out new leadership strategies, you can benefit yourself and others by sharing these experiences with your peers. This may look like literally passing this book on to a friend. Or it could mean meeting up with a colleague and asking how they handle leadership situations like conflict resolution or goal setting. Unlike some things in academia, leadership is not a zero-sum game. When we support each other in improving our leadership skills and work together to become better at the people-facing side of this job, we all win. And the people who win the most are those who are earliest in their careers and are counting on us to get it right.

# NOTES

# ACKNOWLEDGMENTS

# INDEX

# NOTES

## 1. Time Management

1. Hal G. Ersner-Hershfield, Elliott Wimmer, and Brian Knutson, "Saving for the Future Self: Neural Measures of Future Self-Continuity Predict Temporal Discounting," *Journal of Social Cognitive Affective Neuroscience* 4, no. 1 (2009): 85–92.
2. Ersner-Hershfield, Wimmer, and Knutson, "Saving for the Future Self"; Gretchen B. Chapman, "Temporal Discounting and Utility for Health and Money," *Journal of Experimental Psychology: Learning, Memory, and Cognition* 22, no. 3 (1996): 771–791.
3. Hal G. Ersner-Hershfield et al., "Don't Stop Thinking about Tomorrow: Individual Differences in Future Self-Continuity Account for Saving," *Judgment and Decision Making* 4, no. 4 (2009): 280–286; Abraham M. Rutchick et al., "Future Self-Continuity Is Associated with Improved Health and Increases Exercise Behavior," *Journal of Experimental Psychology: Applied* 24, no. 1 (2018): 72–80.
4. Rutchick et al., "Future Self-Continuity," 72.
5. Amy Wrzesniewski, Justin M. Berg, and Jane E. Dutton, "Turn the Job You Have into the Job You Want," *Harvard Business Review* 88, no. 6 (2010): 114–117.

6. Tiffany Dufu, *Drop the Ball: Achieving More by Doing Less* (New York: Flatiron Books, 2017).
7. Lindsey Dugdill et al., "Exercising at Work and Self-Reported Work Performance," *International Journal of Workplace Health Management* 1, no. 3 (2008): 176–197.
8. M. Küüsmaa-Schildt et al., "Effects of Morning vs. Evening Combined Strength and Endurance Training on Physical Performance, Sleep and Well-Being," *Chronobiology International* 36, no. 6 (2019): 811–825.
9. Cheryl J. Hansen, Larry C. Stevens, and J. Richard Coast, "Exercise Duration and Mood State: How Much Is Enough to Feel Better?" *Health Psychology* 20, no. 4 (2001): 267; Chi Pang Wen et al., "Minimal Amount of Exercise to Prolong Life: To Walk, to Run, or Just Mix It Up?" *Journal of the American College of Cardiology* 64, no. 5 (2014): 482–484.

## 2. Leadership Strengths and Styles

1. Marcus Buckingham, with Adam Grant, host, *WorkLife with Adam Grant*, podcast, "When Strength Becomes Weakness," April 24, 2019, 36:42.
2. Deniz Ones and Stephan Dilchert, "How Special Are Executives? How Special Should Executive Selection Be? Observations and Recommendations." *Industrial and Organizational Psychology* 2, no. 2 (2009): 163–170; Lindsay G. Lebin et al., "Continuing the Quiet Revolution: Developing Introverted Leaders in Academic Psychiatry," *Academic Psychiatry* 43 (2019): 516–520; Adam M. Grant, Francesca Gino, and David A. Hofmann, "The Hidden Advantages of Quiet Bosses," *Harvard Business Review* 88, no. 12 (2010): 28.
3. John H. Zenger and Joseph R. Folkman, *The Extraordinary Leader: Turning Good Managers into Great Leaders* (New York: McGraw-Hill, 2009).
4. Jack Zenger and Joseph Folkman, "Ten Fatal Flaws That Derail Leaders," *Harvard Business Review* 87, no. 6 (2009): 18.

5. Wiebke Bleidorn et al., "Personality Trait Stability and Change," *Personality Science* 2 (2021): 1–20.
6. Meera Komarraju et al., "The Big Five Personality Traits, Learning Styles, and Academic Achievement," *Personality and Individual Differences* 51, no. 4 (2011): 472–477; Sarah Brown and Karl Taylor, "Household Finances and the 'Big Five' Personality Traits," *Journal of Economic Psychology* 45 (2014): 197–212; Danny Azucar, Davide Marengo, and Michele Settanni, "Predicting the Big 5 Personality Traits from Digital Footprints on Social Media: A Meta-Analysis," *Personality and Individual Differences* 124 (2018): 150–159.
7. Anna Sutton, Chris Allinson, and Helen Williams, "Personality Type and Work-Related Outcomes: An Exploratory Application of the Enneagram Model," *European Management Journal* 31, no. 3 (2013): 234–249.
8. Marcus Buckingham and Donald O. Clifton, *Now, Discover Your Strengths* (New York: Free Press, 2001). "CliftonStrengths" is a registered trademark.
9. Sharon Birkman Fink and Stephanie Capparell, *The Birkman Method: Your Personality at Work* (San Francisco: Jossey-Bass, 2013). "Birkman Method" is a registered trademark.
10. Daniel Goleman, *Leadership That Gets Results* (Boston: Harvard Business Review Press, 2017).

## 3. Goals

1. Robert E. Wood, "Task Complexity: Definition of the Construct," *Organizational Behavior and Human Decision Processes* 37, no. 1 (1986): 60–82.
2. George T. Doran, "There's a Smart Way to Write Management's Goals and Objectives," *Management Review* 70, no. 11 (1981): 35–36. Edwin A. Locke and Gary P. Latham, "Building a Practically Useful Theory of Goal Setting and Task Motivation: A 35-Year Odyssey," *American Psychologist* 57, no. 9 (2002): 705–717.

3. Peter M. Gollwitzer and Paschal Sheeran, "Implementation Intentions and Goal Achievement: A Meta-Analysis of Effects and Processes," *Advances in Experimental Social Psychology* 38 (2006): 69–119.
4. Tracy Epton, Sinead Currie, and Christopher J. Armitage, "Unique Effects of Setting Goals on Behavior Change: Systematic Review and Meta-Analysis," *Journal of Consulting and Clinical Psychology* 85, no. 12 (2017): 1182–1198.
5. Henk Aarts, Peter M. Gollwitzer, and Ran R. Hassin, "Goal Contagion: Perceiving Is for Pursuing," *Journal of Personality and Social Psychology* 87, no. 1 (2004): 23–37.
6. Chris Loersch et al., "The Influence of Social Groups on Goal Contagion," *Journal of Experimental Social Psychology* 44, no. 6 (2008): 1555–1558.
7. Shai Davidai and Sebastian Deri, "The Second Pugilist's Plight: Why People Believe They Are above Average but Are Not Especially Happy about It," *Journal of Experimental Psychology: General* 148, no. 3 (2019): 570–587.

# 4. Motivation

1. Joseph R. Ferrari et al., "The Antecedents and Consequences of Academic Excuse-Making: Examining Individual Differences in Procrastination," *Research in Higher Education* 39, no. 2 (1998): 199–215; Joseph R. Ferrari, "Self-Handicapping by Procrastinators: Protecting Self-Esteem, Social-Esteem, or Both?," *Journal of Research in Personality* 25, no. 3 (1991): 245–261.
2. Gregory Schraw, Theresa Wadkins, and Lori Olafson, "Doing the Things We Do: A Grounded Theory of Academic Procrastination," *Journal of Educational Psychology* 99, no. 1 (2007): 12–25.
3. Angela Hsin Chun Chu and Jin Nam Choi, "Rethinking Procrastination: Positive Effects of 'Active' Procrastination Behavior on Attitudes and Performance," *Journal of Social Psychology* 145, no. 3 (2005): 245–264.

4. Richard M. Ryan and Edward L. Deci, "Intrinsic and Extrinsic Motivations: Classic Definitions and New Directions," *Contemporary Educational Psychology* 25, no. 1 (2000): 54–67.
5. Teresa M. Amabile, "Motivation and Creativity: Effects of Motivational Orientation on Creative Writers," *Journal of Personality and Social Psychology* 48, no. 2 (1985): 393–399; Bård Kuvaas et al., "Do Intrinsic and Extrinsic Motivation Relate Differently to Employee Outcomes?," *Journal of Economic Psychology* 61 (2017): 244–258.
6. Christopher T. Pisarik, "Motivational Orientation and Burnout among Undergraduate College Students," *College Student Journal* 43, no. 4 (2009): 1238–1253; Scott L. Cresswell and Robert C. Eklund, "Motivation and Burnout in Professional Rugby Players," *Research Quarterly for Exercise and Sport* 76, no. 3 (2005): 370–376.
7. Giovanni B. Moneta, "Opportunity for Creativity in the Job as a Moderator of the Relation between Trait Intrinsic Motivation and Flow in Work," *Motivation and Emotion* 36, no. 4 (2012): 491–450.
8. Ryan and Deci, "Intrinsic and Extrinsic Motivations."
9. Daniel H. Pink, *Drive: The Surprising Truth about What Motivates Us* (New York: Riverhead Books, 2009).
10. T. Cameron Wild and Michael E. Enzle, "Social Contagion of Motivational Orientations," in *Handbook of Self-Determination Research*, ed. Edward L. Deci and Richard M. Ryan (Rochester, NY: University of Rochester Press, 2002), 141–157.

# 5. Resilience

1. "BU Chemistry Professor Malika Jeffries-EL '96 Encourages Wellesley Students to Persevere in the Lab and in Life," Wellesley College, News, September 2, 2016, www1.wellesley.edu/news/2016/september/node/97721.

2. David E. Conroy, "Progress in the Development of a Multidimensional Measure of Fear of Failure: The Performance Failure Appraisal Inventory (PFAI)," *Anxiety, Stress, and Coping* 14, no. 4 (2001): 431–452.
3. Carol S. Dweck, "Implicit Theories as Organizers of Goals and Behavior," in *The Psychology of Action: Linking Cognition and Motivation to Behavior*, ed. Peter M. Gollwitzer and John A. Bargh (New York: Guilford Press, 1996).
4. Carol S. Dweck, *Mindset: The New Psychology of Success* (New York: Ballantine, 2006).
5. Andrew J. Elliot and Marcy A. Church, "A Motivational Analysis of Defensive Pessimism and Self-Handicapping," *Journal of Personality* 71, no. 3 (2003): 369–396.
6. Charles S. Carver et al., "Assessing Coping Strategies: A Theoretically Based Approach," *Journal of Personality and Social Psychology* 56, no. 2 (1989): 267–283.
7. Paul Rozin and Edward B. Royzman, "Negativity Bias, Negativity Dominance, and Contagion," *Personality and Social Psychology Review* 5, no. 4 (2001): 296–320; Ian H. Gotlib, "Perception and Recall of Interpersonal Feedback: Negative Bias in Depression," *Cognitive Therapy and Research* 7, no. 5 (1983): 399–412.
8. Eric Robinson et al., "Eating Attentively: A Systematic Review and Meta-Analysis of the Effect of Food Intake Memory and Awareness on Eating," *American Journal of Clinical Nutrition* 97, no. 4 (2013): 728–742.
9. Pauline Rose Clance and Suzanne Ament Imes, "The Imposter Phenomenon in High Achieving Women: Dynamics and Therapeutic Intervention," *Psychotherapy: Theory, Research & Practice* 15, no. 3 (1978): 241–247.

## 6. Receiving Feedback

1. Alex Forsythe and Sophie Johnson, "Thanks, but No-Thanks for the Feedback," *Assessment & Evaluation in Higher Education* 42, no. 6 (2017): 850–859.

2. Olle Th. J. ten Cate, "Why Receiving Feedback Collides with Self Determination," *Advances in Health Sciences Education* 18, no. 4 (2013): 845–849.
3. David E. Conroy, "Progress in the Development of a Multidimensional Measure of Fear of Failure: The Performance Failure Appraisal Inventory (PFAI)," *Anxiety, Stress, and Coping* 14, no. 4 (2001): 431–452.
4. Amy C. Edmondson, "Speaking Up in the Operating Room: How Team Leaders Promote Learning in Interdisciplinary Action Teams," *Journal of Management Studies* 40, no. 6 (2003): 1419–1452.
5. Edgar H. Schein and Warren G. Bennis, *Personal and Organizational Change through Group Methods: The Laboratory Approach* (New York: Wiley, 1965).
6. Robert T. Golembiewski, ed., *Handbook of Organizational Behavior*, 2nd ed. (New York: Marcel Dekker, 2001).
7. Edmondson, "Speaking Up."

# 7. Ownership

1. Julian Gould-Williams, "The Importance of HR Practices and Workplace Trust in Achieving Superior Performance: A Study of Public-Sector Organizations," *International Journal of Human Resource Management* 14, no. 1 (2003): 28–54; Sarah Brown, "Employee Trust and Workplace Performance," *Journal of Economic Behavior & Organization* 116 (2015): 361–378.
2. Louise Young and Kerry Daniel, "Affectual Trust in the Workplace," *International Journal of Human Resource Management* 14, no. 1 (2003): 139–155; Roderick M. Kramer, "Trust and Distrust in Organizations: Emerging Perspectives, Enduring Questions," *Annual Review of Psychology* 50, no. 1 (1999): 569–598.
3. Amy C. Edmondson, "Psychological Safety, Trust, and Learning in Organizations: A Group-Level Lens," in *Trust and Distrust in Organizations: Dilemmas and Approaches*, ed. Roderick M. Kramer and Karen S. Cook (New York: Russell Sage Foundation, 2004), 239–272.
4. Young and Daniel, "Affectual Trust."

## 8. Environment

1. Examples of antiracism and DEI resource lists can be found at: https://www.bu.edu/alumni/2022/06/10/blog-juneteenth-reading-list-antiracism/; https://guides.library.uwm.edu/lists; https://caes.ucdavis.edu/about/leaders/DEI/books.
2. Joseph E. McGrath, Jennifer L Berdahl, and Holly Arrow, "Traits, Expectations, Culture, and Clout: The Dynamics of Diversity in Work Groups," in *Diversity in Work Teams: Research Paradigms for a Changing Workplace*, ed. Susan E. Jackson and Marian N. Ruderman (Washington, DC: American Psychological Association, 1995).
3. David Rock, Heidi Grant, and Jacqui Grey, "Diverse Teams Feel Less Comfortable—and That's Why They Perform Better," *Harvard Business Review* 95, no. 9 (2016): 22.
4. Christine Yifeng Chen et al., "Meta-Research: Systemic Racial Disparities in Funding Rates at the National Science Foundation," *ELife* 11 (2022): e83071; Donna K. Ginther et al., "Race, Ethnicity, and NIH Research Awards," *Science* 333, no. 6045 (2011): 1015–1019; Susan W. White, Mengya Xia, and Gabrielle Edwards, "Race, Gender, and Scholarly Impact: Disparities for Women and Faculty of Color in Clinical Psychology," *Journal of Clinical Psychology* 77, no. 1 (2021): 78–89; Jennifer M. Kresbach, "Women in Academia: Representation, Tenure, and Publication Patterns in the STEM and Social Sciences Fields," *Journal of International Women's Studies* 24, no. 5 (2022): 1–15; Jevin D. West et al., "The Role of Gender in Scholarly Authorship," *PLoS One* 8, no. 7 (2013): e66212; Casey Miller and Keivan Stassun, "A Test That Fails," *Nature* 510, no. 7504 (2014): 303–304; Liane Moneta-Koehler et al., "The Limitations of the GRE in Predicting Success in Biomedical Graduate School," *PLoS One* 12, no. 1 (2017): e0166742.
5. Kimberlé Crenshaw, "Demarginalizing the Intersection of Race and Sex: A Black Feminist Critique of Antidiscrimination Doctrine, Feminist

Theory and Antiracist Politics," in *Feminist Legal Theories: Readings in Law and Gender*, ed. and introduction by Karen J. Maschke, Gender and American Law (New York: Routledge, 2013).

6. Nidhi Subbaraman, "How #BlackInTheIvory Put a Spotlight on Racism in Academia," *Nature* 582, no. 7812 (2020): 327–328.

7. Anthony Abraham Jack, *The Privileged Poor: How Elite Colleges Are Failing Disadvantaged Students* (Cambridge, MA: Harvard University Press, 2019).

8. Daniel J. Brown et al., "Human Thriving," *European Psychologist* 22 (2017): 167–179.

9. Tamra C. Blue, "Beyond a Seat at the Table," *Nature Reviews Chemistry* 6, no. 4 (2022): 235–236.

10. Terrell R. Morton, "A Phenomenological and Ecological Perspective on the Influence of Undergraduate Research Experiences on Black Women's Persistence in STEM at an HBCU," *Journal of Diversity in Higher Education* 14, no. 4 (2020): 530–543.

11. Mica Estrada, Alegra Eroy-Reveles, and John Matsui, "The Influence of Affirming Kindness and Community on Broadening Participation in STEM Career Pathways," *Social Issues and Policy Review* 12, no. 1 (2018): 258–279.

12. Juanita C. Limas et al., "The Impact of Research Culture on Mental Health & Diversity in STEM," *Chemistry: A European Journal* 28 (2022): e202102957.

13. Laura Weiss Roberts, "Belonging, Respectful Inclusion, and Diversity in Medical Education," *Academic Medicine* 95, no. 5 (2020): 661–664.

## 9. Giving Feedback

1. D. Royce Sadler, "Formative Assessment and the Design of Instructional Systems," *Instructional Science* 18, no. 2 (1989): 119–144.

2. Alice Man Sze Lau, "'Formative Good, Summative Bad?' A Review of the Dichotomy in Assessment Literature," *Journal of Further Higher Education* 40, no. 4 (2016): 509–525.
3. Robert Rosenthal and Lenore Jacobson, "Pygmalion in the Classroom," *Urban Review* 3, no. 1 (1968): 16–20.
4. Christopher DeLuca, Andrew Coombs, and Danielle LaPointe-McEwan, "Assessment Mindset: Exploring the Relationship between Teacher Mindset and Approaches to Classroom Assessment," *Studies in Educational Evaluation* 61 (2019): 159–169.
5. Ed Catmull with Amy Wallace, *Creativity, Inc.: Overcoming the Unseen Forces That Stand in the Way of True Inspiration,* expanded ed., 2nd ed. (New York: Random House, 2023).
6. Catmull and Wallace, *Creativity, Inc.,* 2.
7. Kim Scott, *Radical Candor: Be a Kick-Ass Boss without Losing Your Humanity* (New York: St. Martin's Press, 2017).
8. Karen Kirkland and Sam Manoogian, *Ongoing Feedback: How to Get It, How to Use It* (Greensboro, NC: Center for Creative Leadership, 1998).
9. Michael P. O'Driscoll and Terry A. Beehr, "Supervisor Behaviors, Role Stressors and Uncertainty as Predictors of Personal Outcomes for Subordinates," *Journal of Organizational Behavior* 15, no. 2 (1994): 141–155.
10. Marcial Losada and Emily Heaphy, "The Role of Positivity and Connectivity in the Performance of Business Teams: A Nonlinear Dynamics Model," *American Behavioral Scientist* 47, no. 6 (2004): 740–765; Lia Voerman et al., "Types and Frequencies of Feedback Interventions in Classroom Interaction in Secondary Education," *Teaching and Teacher Education* 28, no. 8 (2012): 1107–1115.

## 10. Conflict Resolution

1. Leslie A. DeChurch and Michelle A. Marks, "Maximizing the Benefits of Task Conflict: The Role of Conflict Management," *International Journal*

of *Conflict Management* 12, no. 1 (2001): 4–22; Bret H. Bradley et al., "Reaping the Benefits of Task Conflict in Teams: The Critical Role of Team Psychological Safety Climate," *Journal of Applied Psychology* 97, no. 1 (2012): 151–158.

2. Carsten K. W. De Dreu, "The Virtue and Vice of Workplace Conflict: Food for (Pessimistic) Thought," *Journal of Organizational Behavior* 29, no. 1 (2008): 5–18.

3. Joseph Grenny et al., *Crucial Conversations: Tools for Talking When Stakes Are High*, 3rd ed. (New York: McGraw-Hill, 2021).

4. Karina S. Blair et al., "Modulation of Emotion by Cognition and Cognition by Emotion," *Neuroimage* 35, no. 1 (2007): 430–440.

5. Fabiana Silva Ribeiro et al., "How Does Allocation of Emotional Stimuli Impact Working Memory Tasks? An Overview," *Advances in Cognitive Psychology* 15, no. 2 (2019): 155–168.

6. Roger Fisher, William L. Ury, and Bruce Patton, *Getting to Yes: Negotiating Agreement without Giving In*, 3rd ed. (New York: Penguin, 2011).

7. Sheila Heen, with Adam Grant, host, *WorkLife with Adam Grant*, podcast, "The Office without A**holes," April 2, 2019, 41:53.

8. Randy A. Sansone and Lori A. Sansone, "Personality Disorders: A Nation-Based Perspective on Prevalence," *Innovations in Clinical Neuroscience* 8, no. 4 (2011): 13–18.

9. Bridget F. Grant et al., "Co-occurrence of 12-Month Alcohol and Drug Use Disorders and Personality Disorders in the United States: Results from the National Epidemiologic Survey on Alcohol and Related Conditions," *Archives of General Psychiatry* 61, no. 4 (2004): 361–368; Frederick S. Stinson et al., "Prevalence, Correlates, Disability, and Comorbidity of DSM-IV Narcissistic Personality Disorder: Results from the Wave 2 National Epidemiologic Survey on Alcohol and Related Conditions," *Journal of Clinical Psychiatry* 69, no. 7 (2008): 1033–1045; Bridget F. Grant et al., "Prevalence, Correlates, Disability, and Comorbidity of DSM-IV Border-

line Personality Disorder: Results from the Wave 2 National Epidemiologic Survey on Alcohol and Related Conditions," *Journal of Clinical Psychiatry* 69, no. 4 (2008): 533–545.
10. Alice Diedrich and Ulrich Voderholzer, "Obsessive-Compulsive Personality Disorder: A Current Review," *Current Psychiatry Reports* 17 (2015): 1–10.
11. John G. Gunderson et al., "Borderline Personality Disorder," *Nature Reviews: Disease Primers* 4, no. 1 (2018): 1–20.
12. Aaron L. Pincus and Mark R. Lukowitsky, "Pathological Narcissism and Narcissistic Personality Disorder," *Annual Review of Clinical Psychology* 6, no. 1 (2010): 421–446.
13. Kathrin Ritter et al., "Lack of Empathy in Patients with Narcissistic Personality Disorder," *Psychiatry Research* 187, no. 1–2 (2011): 241–247; Eve Caligor, Kenneth N. Levy, and Frank E. Yeomans, "Narcissistic Personality Disorder: Diagnostic and Clinical Challenges," *American Journal of Psychiatry* 172, no. 5 (2015): 415–422.

## 11. Ethical Leadership

1. Michael E. Brown, Linda K. Treviño, and David A. Harrison, "Ethical Leadership: A Social Learning Perspective for Construct Development and Testing," *Organizational Behavior and Human Decision Processes* 97, no. 2 (2005): 117–134.
2. Gary Yukl et al., "An Improved Measure of Ethical Leadership," *Journal of Leadership and Organizational Studies* 20, no. 1 (2013): 38–48; David M. Mayer, Maribeth Kuenzi, and Rebecca L. Greenbaum, "Examining the Link between Ethical Leadership and Employee Misconduct: The Mediating Role of Ethical Climate," *Journal of Business Ethics* 95, no. 1 (2010): 7–16.
3. Max H. Bazerman and Ann E. Tenbrunsel, "Ethical Breakdowns," *Harvard Business Review* 89, no. 4 (April 2011): 58–137.

4. John H. Zenger and Joseph R. Folkman, *The Extraordinary Leader: Turning Good Managers into Great Leaders* (New York: McGraw-Hill, 2009).
5. See Jelena Morris and Janet Holt, "Applying Utilitarianism to the Presumed Consent System for Organ Donation to Consider the Moral Pros and Cons," *British Journal of Nursing* 30, no. 19 (2021): 1127–1131.
6. Alison L. Antes, Ashley Kuykendall, and James M. DuBois, "The Lab Management Practices of 'Research Exemplars' That Foster Research Rigor and Regulatory Compliance: A Qualitative Study of Successful Principal Investigators," *PLoS One* 14, no. 4 (2019): e0214595.
7. Tristan McIntosh, Alison L. Antes, and James M. DuBois, "Navigating Complex, Ethical Problems in Professional Life: A Guide to Teaching Smart Strategies for Decision-Making," *Journal of Academic Ethics* 19, no. 2 (2021): 139–156.
8. Max H. Bazerman, "A New Model for Ethical Leadership," *Harvard Business Review* 8, no. 5 (September–October 2020): 90–97.

## 12. Communication

1. Laura Stafford and John A. Daly, "Conversational Memory: The Effects of Recall Mode and Memory Expectancies on Remembrances of Natural Conversations," *Human Communication Research* 10, no. 3 (1984): 379–402.
2. Gretchen R. Vogelgesang, Hannes Leroy, and Bruce J. Avolio, "The Mediating Effects of Leader Integrity with Transparency in Communication and Work Engagement/Performance," *Leadership Quarterly* 24, no. 3 (2013): 405–413.
3. Mariale Hardiman, "Informing Pedagogy through the Brain-Targeted Teaching Model," *Journal of Microbiology Biology Education* 13, no. 1 (2012): 11–16.
4. Lars Schwabe and Oliver T. Wolf, "Learning under Stress Impairs Memory Formation," *Neurobiology of Learning and Memory* 93, no. 2 (2010): 183–188.

5. Janet E. Resop Reilly and Joyce J. Fitzpatrick, "Perceived Stress and Sense of Belonging in Doctor of Nursing Practice Students," *Journal of Professional Nursing* 25, no. 2 (2009): 81–86.
6. Annikka Chrestensen, "Real-World Context, Interest, Understanding, and Retention" (master's thesis, Michigan Technological University, 2007).
7. Peter S. Houts et al., "The Role of Pictures in Improving Health Communication: A Review of Research on Attention, Comprehension, Recall, and Adherence," *Patient Education and Counseling* 61, no. 2 (2006): 173–190.
8. Marshall A. Baker and J. Shane Robinson, "The Effect of Two Different Pedagogical Delivery Methods on Students' Retention of Knowledge over Time," *Journal of Agricultural Education* 59, no. 1 (2018): 100–118.
9. Steven M. Smith, Arthur Glenberg, and Robert A. Bjork, "Environmental Context and Human Memory," *Memory & Cognition* 6, no. 4 (1978): 342–353.
10. Benjamin C. Storm, "The Benefit of Forgetting in Thinking and Remembering," *Current Directions in Psychological Science* 20, no. 5 (2011): 291–295.
11. Daniel T. Willingham, Elizabeth M. Hughes, and David G. Dobolyi, "The Scientific Status of Learning Styles Theories," *Teaching of Psychology* 42, no. 3 (2015): 266–271.

## 13. Pass It On

1. Martin Stigmar, "Peer-to-Peer Teaching in Higher Education: A Critical Literature Review," *Mentoring & Tutoring* 24, no. 2 (2016): 124–136.
2. Robert Rosenthal and Lenore Jacobson, "Pygmalion in the Classroom," *Urban Review* 3, no. 1 (1968): 16–20.
3. Diane Wood Allen, "How Nurses Become Leaders: Perceptions and Beliefs about Leadership Development," *Journal of Nursing Administration* 28, no. 9 (1998): 15–20.

4. Daniela Rodrigues, Cristina Padez, and Aristides M. Machado-Rodrigues, "Active Parents, Active Children: The Importance of Parental Organized Physical Activity in Children's Extracurricular Sport Participation," *Journal of Child Health Care* 22, no. 1 (2018): 159–170; Helene Raskin White, Valerie Johnson, and Steven Buyske, "Parental Modeling and Parenting Behavior Effects on Offspring Alcohol and Cigarette Use: A Growth Curve Analysis," *Journal of Substance Abuse* 12, no. 3 (2000): 287–310.
5. Michael E. Brown and Linda K. Treviño, "Do Role Models Matter? An Investigation of Role Modeling as an Antecedent of Perceived Ethical Leadership," *Journal of Business Ethics* 122 (2014): 587–598.
6. David Duran, "Learning-by-Teaching: Evidence and Implications as a Pedagogical Mechanism," *Innovations in Education Teaching International* 54, no. 5 (2017): 476–484.

# 14. Vulnerability

1. Richard Farson and Ralph Keyes, "The Failure-Tolerant Leader," *Harvard Business Review* 80, no. 8 (2002): 64–71.
2. Brené Brown, *Daring Greatly: How the Courage to Be Vulnerable Transforms the Way We Live, Love, Parent, and Lead* (New York: Avery, an imprint of Penguin Random House, 2015).
3. Crystal L. Hoyt, Jeni L. Burnette, and Audrey N. Innella, "I Can Do That: The Impact of Implicit Theories on Leadership Role Model Effectiveness," *Personality and Social Psychology Bulletin* 38, no. 2 (2012): 257–268.
4. Holly M. Hutchins, "Outing the Imposter: A Study Exploring Imposter Phenomenon among Higher Education Faculty," *New Horizons in Adult Education and Human Resource Development* 27, no. 2 (2015): 3–12.
5. Anna Parkman, "The Imposter Phenomenon in Higher Education: Incidence and Impact," *Journal of Higher Education Theory and Practice* 16, no. 1 (2016): 51–60.

6. Dena M. Bravata et al., "Prevalence, Predictors, and Treatment of Impostor Syndrome: A Systematic Review," *Journal of General Internal Medicine* 35 (2020): 1252–1275.
7. "Ex-Lockheed Chairman Daniel Haughton, 75, Dies," *Los Angeles Times*, July 6, 1987.
8. Victoria L. Brescoll, Erica Dawson, and Eric Luis Uhlmann, "Hard Won and Easily Lost: The Fragile Status of Leaders in Gender-Stereotype-Incongruent Occupations," *Psychological Science* 21, no. 11 (2010): 1640–1642.
9. Ashleigh Shelby Rosette and Robert W. Livingston, "Failure Is Not an Option for Black Women: Effects of Organizational Performance on Leaders with Single versus Dual-Subordinate Identities," *Journal of Experimental Social Psychology* 48, no. 5 (2012): 1162–1167.
10. Jody Condit Fagan and Ari Emilia Short, "Searching for Nonbinary and Trans Inclusion: A Call to Action for Leadership Studies," *Libraries* (2023): 228.
11. Anna M. Agathangelou and Lily H. M. Ling, "An Unten(ur)able Position: The Politics of Teaching for Women of Color in the US," *International Feminist Journal of Politics* 4, no. 3 (2002): 368–398.
12. Robin Selzer, Amy Howton, and Felicia Wallace, "Rethinking Women's Leadership Development: Voices from the Trenches," *Administrative Sciences* 7, no. 2 (2017): 18.
13. Brené Brown, *Dare to Lead: Brave Work, Tough Conversations, Whole Hearts* (New York: Random House, 2018).

## 15. Challenging Situations

1. Fredrik Bondestam and Maja Lundqvist, "Sexual Harassment in Higher Education—A Systematic Review," *European Journal of Higher Education* 10, no. 4 (2020): 397–419; Naif Fnais et al., "Harassment and Discrimination in Medical Training: A Systematic Review and Meta-Analysis," *Aca-

demic Medicine 89, no. 5 (2014): 817–827; Emily A. Vargas et al., "Incidence and Group Comparisons of Harassment Based on Gender, LGBTQ+ Identity, and Race at an Academic Medical Center," Journal of Women's Health 30, no. 6 (2021): 789–798.
2. Isis H. Settles et al., "The Climate for Women in Academic Science: The Good, the Bad, and the Changeable," Psychology of Women Quarterly 30, no. 1 (2006): 47–58.
3. Steven M. Norman, Bruce J. Avolio, and Fred Luthans, "The Impact of Positivity and Transparency on Trust in Leaders and Their Perceived Effectiveness," Leadership Quarterly 21, no. 3 (2010): 350–364.
4. Ruth McKay et al., "Workplace Bullying in Academia: A Canadian Study," Employee Responsibilities and Rights Journal 20, no. 2 (2008): 77–100.
5. Joseph Grenny et al., Crucial Conversations: Tools for Talking When Stakes Are High, 3rd ed. (New York: McGraw-Hill, 2021).

## 16. Empowering

1. Jean Lau Chin, "Diversity Leadership: Influence of Ethnicity, Gender, and Minority Status," Open Journal of Leadership 2, no. 1 (2013): 1; Gaëtane Jean-Marie, Vicki A. Williams, and Sheila L. Sherman, "Black Women's Leadership Experiences: Examining the Intersectionality of Race and Gender," Advances in Developing Human Resources 11, no. 5 (2009): 562–581.
2. Brenda L. McKenzie, "Am I a Leader? Female Students Leadership Identity Development," Journal of Leadership Education 17, no. 2 (2018).
3. Karyn Wulf et al., "Personality Compatibility within Faculty Mentoring Dyads and Perceived Mentoring Outcomes: Survey Results of Academic Medicine Institutions in the USA," Medical Science Educator 31, no. 2 (2021): 345–348; Christine Menges, "Toward Improving the Effectiveness of Formal Mentoring Programs: Matching by Personality Matters," Group Organization Management 41, no. 1 (2016): 98–129.

4. Cynthia D. McCauley and Ellen Van Velsor, eds., *The Center for Creative Leadership Handbook of Leadership Development*, 2nd ed. (Hoboken, NJ: Jossey-Bass, 2003).

5. Paul J. H. Schoemaker, Steve Krupp, and Samantha Howland, "Strategic Leadership: The Essential Skills," *Harvard Business Review* 91, no. 1 (2013): 131–134.

# ACKNOWLEDGMENTS

Chemistry professors don't typically write leadership books. Thus, the fact that this book exists is a testament to the amazing village of people in my life who spoke this wild idea into existence, kept my fear and self-doubt from shutting me down, inspired and improved the content and delivery, and walked by my side to carry the project across the finish line.

I have to begin by acknowledging my research group, as that is where this project began. Thank you to Shayla Shorter for the number of times you said, "Jen, you should write a book" and then, when I responded with "I could never do that," gave me a kind yet firm look that said "that doesn't sound like a growth mindset." I'm also extraordinarily grateful to all of the members of the Heemstra Lab—past, present, and future—for embarking with me on this leadership journey that I didn't expect and wasn't prepared for. Thank you to those who gracefully bore with me in my first attempt at conflict resolution, didn't lose hope when our grant proposals were rejected, stood together through a bad tenure vote and two subsequent moves, and have been fantastic colleagues through all of the ups and downs of academic research life. I can guarantee that you have taught me more about leadership and mentoring than I could ever teach you about chemistry, and I treasure the ways that you motivate me every day to keep learning and growing.

## ACKNOWLEDGMENTS

I'm similarly grateful to those who have directly mentored me as a member of their research groups—James Nowick, Jeff Moore, and David Liu. Thanks to each of you for providing such a positive model of what mentorship and leadership should look like in a research lab and for your consistent support and encouragement, which have been critical to my career. It has been a privilege to work with you and to see our relationships evolve from mentor-mentee to that of peers and friends. Special thanks to Jeff for his permission to share some of our conversations in this book and for helping me see opportunities in leadership that I didn't think could exist.

I also want to thank those in my research community, social media network, and beyond (too many to name here!), who have graciously offered their leadership wisdom over the years and been a sounding board for working through ideas and challenges. Thank you for being open with your own struggles and helping me to see that I'm not alone in the feeling of "I wasn't trained for this." Special thanks to my dear friend Troy Champ for introducing me to the field of organizational leadership and for asking me the question that changed my approach to time management.

Thank you to everyone who has coached me through this project (even when you didn't realize you were coaching me). I'm especially grateful to Amanda Shaffer for the knowledge and wisdom she has given me over the years that is now embedded in stories throughout this book. I'll never forget saying in one of our coaching sessions, "If I do decide to write a book—" and having you immediately cut in with a response along the lines of, "Is that something that you're really still deciding? You may not have decided *when* you will write a book, but I think you've decided *that* you will write a book." Amanda, you had no idea how much I needed to hear those words in that moment. (Actually, you probably did know because you're an amazing coach!) Thanks also to Ken Carter, Amy Edmondson, Peggy Guest, Matt Hartings, John Inazu, Ellen Levine, and Anne Trubek for their coaching and encouragement throughout the writing and publishing process.

## ACKNOWLEDGMENTS

I also want to thank my friends and colleagues, including Alison Antes, Lou Charkoudian, Stanna Dorn, Mica Estrada, Katherine Haas, Felix Kaspar, Tamra Lahom, Will Pomerantz, and Kira Walsh, who generously volunteered their time to read drafts of the proposal, chapters, and complete manuscript, and who offered candid, incisive, and supportive feedback. I'm grateful for how you sharpened my writing, pushed me to support claims with evidence, and helped me to capture as many perspectives as possible in these pages. Special thanks to Lou for not only reading (and in some cases rereading) every chapter, but also being a constant source of leadership wisdom and personal encouragement throughout this project. I'm also very grateful to the individuals I can name only as Reviewer 1 and Reviewer 2. I may never know your identities, but I want you to know that your constructive and encouraging comments made this a better book—thank you.

When I set out to write *Labwork to Leadership*, I had no idea how to even find a publisher, let alone how a simple Word document on my computer would be transformed into a book that someone could hold in their hands or share with a friend. Thank you to all of the editors and editorial staff who have made this possible. This includes Sara Tenney, Linda Wang, and Amanda Yarnell, who introduced me to the world of academic publishing; Janice Audet, who saw and believed in the vision for this project, provided the opportunity to publish with Harvard University Press, and sharpened both the chapters themselves and the overall arc of the book; and Rachel Field, Aaron Wistar, Kate Brick, and the entire team at Harvard University Press for providing feedback and support that were critical in bringing this book to completion.

Finally, and most important, I am extraordinarily grateful to my friends and family who have walked together with me on this adventure. When I started outlining this book in 2019, I had no idea that my writing trajectory would be intersected by a global pandemic and a major career transition. Each of the times that you (nonjudgmentally) asked, "How is your book coming along?" provided the fuel that got me across the finish line. I'm especially grateful to the three people who provided this support on a daily basis. To my kids—Evan and Owen—thank you for being in my life, for being the kind and creative individuals you are, and for the good-morning hugs and smiles that make each day better, especially when I'm struggling with early-morning writer's block. To my spouse and the love of my life, John, I couldn't possibly put all of my gratitude into words here. I'm thankful for everything, from your asking the simple question, "What would it look like to try?" when I mentioned writing a book, to the daily leadership wisdom I gain from our conversations, to the million ways that you have supported and encouraged me in my career and in this project. It's been a privilege to navigate through life together since we met as first-year PhD students, and I can't wait for all that the future holds.

For a more complete list of everyone who has made this book possible, please visit my gratitude wall at www.jenheemstra.com/gratitude.

# INDEX

affiliative leadership, 43, 44
affirmation, 137–138
Allen, Diane Wood, 220
"alternative yes," 21
Antes, Alison, 187–188, 190
anxiety. *See* mental health
authoritative leadership, 43, 44
autonomy: feedback and, 95; leadership styles and, 43, 66; motivation and, 69–71, 74, 75; utilitarianism and, 185
Ayres, Zoë, 191

Bailey, Chris, 14
Bazerman, Max, 182, 194
Beehr, Terry, 155
Berg, Justin, 20
Big Five personality test, 35–36, 37
bioethics, 184–185
Birkman Method, 39–42, 66
#BlackInTheIvory, 131
borderline personality disorder (BPD), 175–176
"brain trust" meetings, 150–151
Brown, Brené, 230, 237–238
Brown, Michael, 181, 221
Buckingham, Marcus, 29, 32, 37–38
bullying. *See* harassment, bullying, and discrimination
burnout, 68, 73–74, 233

candor, 144, 150–153, 158
Catmull, Ed, 150–151
Center for Creative Leadership, 154. *See also* Situation-Behavior-Impact
challenging situations, 242–244; action items, 254–255; confidentiality, 244–247, 254–255; conflict example, 249–250; crisis example, 252–253; inappropriate behavior

challenging situations (*continued*)
example, 250–251; performance feedback example, 251–252; types of, 248–253
Champ, Troy, 11
charisma, 31
Choi, Jin Nam, 64
Chu, Angela Hsin Chun, 64
Clifton, Donald, 37–38
CliftonStrengths (StrengthsFinder), 37–39
coaching leadership, 43, 44
coercive leadership, 43, 44
commitment, 272–275; action items, 281–282; communication and, 277–279; evolving roles and, 279–280; lifelong learning and, 280–281; mentoring and, 272–282
communication, 197–199; about communication expectations, 210–212, 213; action items, 212–213; commitment and, 277–279; dissemination versus dialogue, 201–203; information retention and, 206–208; listening skills, 203–206; mental health and, 206–207; repetition and overcommunication, 208–210; transparency in, 199–201
confidentiality, 244–247, 254–255

conflict resolution, 159–161; action items, 178–179; assessing your role in, 161–164, 178–179; BATNA (best alternative to a negotiated agreement), 173; challenging situations, 173–178; desired outcomes of, 167–168; initial investigations, 164–167; mutual purpose and, 168–170, 172–178; mutual respect and, 168, 170–171, 173, 177, 178; negotiation skills and, 172–173; ombudspersons, 163–164, 190; task conflict and, 160; written agreements and records, 171–172
Conroy, David, 80–81, 95
Covey, Stephen, 50
COVID-19 pandemic, 61–62, 205
creativity: communication and, 208–209, 210; in conflict resolution, 160, 175; in DEIJ environments, 130; giving feedback and, 143, 144, 150; leadership strengths and, 18, 31; motivation and, 63, 66, 69, 71, 72
Crenshaw, Kimberlé, 130
Csikszentmihalyi, Mihaly, 63
culture. *See* lab culture

Daly, John, 198
Deci, Edward, 67, 69

# INDEX

decision-making rubric, 17–20
democratic leadership, 43, 44
Dilchert, Stephan, 31
discrimination. *See* harassment, bullying, and discrimination
diversity, equity, inclusion, and justice (DEIJ), 128–131, 263–264; action items, 141–142; assessing, 136–142; culture and, 136–142; environments for, 133–136; equitable work division and, 140; hidden curriculum and, 139–140, 142; intersectionality and, 130–131, 236, 259; mental health and, 140–141, 142; success versus thriving, 131–133; well-being and, 132, 140, 142
Doran, George, 52. *See also* SMART goals
Dufu, Tiffany, 25
Dutton, Jane, 20
Dweck, Carol, 80
Dyson, Brian, 23

Edmondson, Amy, 103–104, 115
email, 22, 205–206, 210–211; appreciative emails, 117; checking, 63; group messaging app versus, 211; in-person dialogue versus, 202; keeping up with, 22; "no" template, 28

empowerment, 256–258; action items, 270–271; graduated learning, 265–268; individuality and, 258–260; mistakes and, 268–270; shared leadership and, 260–265
Enneagram framework, 36–37
environment, DEIJ. *See* diversity, equity, inclusion, and justice
equity. *See* diversity, equity, inclusion, and justice (DEIJ)
Estrada, Mica, 127–128, 137
ethical leadership, 180–182; action items, 195–196; community-driven values and, 185, 186–192; consistency and, 193–195; integrity and, 182, 186, 188, 193–194; mental health and, 191–192, 196; motivated blindness and, 182; multiple positive values, 183–185; personal values and, 185–186, 192–193; research ethics, 185, 186–187, 190, 193, 199, 208; safety and reporting responsibilities, 190–191, 195; utilitarianism and, 184–185
exclusionary experiences, 137
exercise, physical, 23–26
extrinsic motivation, 67–69, 72–73
extroversion, 31, 35, 40

309

# INDEX

failure: coping strategies for, 77, 83–84, 89–90; fear of, 79–82; owning and sharing, 77–79, 89; pain of, 82–84

Farson, Richard, 229–230

feedback, giving, 143–145; action items, 158; candor and kindness in, 150–153, 158; conflicting roles and, 145–146; formative assessment theory, 145–146, 147; formulas for, 153–155; group awards, 157; mental health and, 153, 155; mindset and, 146–148; role of praise in, 156–157, 158; setting expectations for, 149–150, 158; Situation-Behavior-Impact (SBI) model and, 154, 156, 158; for underperformance, 147–149, 152, 153–154

feedback, receiving, 91–93; action items, 108–109; contradictory and poor, 105–108; pain of, 93–96; professional coaching and, 97, 98, 108; psychological safety and, 103–105, 360; feedback exercise, 97–100, 108–109; whom to ask for, 96–99, 108–109

feedback styles, 151–152

Ferrari, Joseph, 62

finances, 172, 173, 207, 226, 230, 235, 264–265

Folkman, Joseph, 32, 182

formative assessment theory, 145–146, 147

future self, letter to, 14

future self-continuity, 13–15

goal contagion, 55–56

goal setting, 47–49; action items, 59–60; comparisons and, 56–59; defining "winning" in, 49–50; driving analogy for, 48–49; five-year goals, 50–52, 54–56, 59; planning, scheduling, and sharing in, 50–55; publication goals, 73, 114, 118–119, 125; retirement party exercise, 50, 59; SMART goals, 52–53, 59, 99

Goleman, Daniel, 42–43, 44

Grant, Adam, 29, 174

grant writing, 265

Gregory, Emily, 167–168, 170

Grenny, Joseph, 167–168, 170

group culture, 4, 31, 192, 239

harassment, bullying, and discrimination, 3; confidentiality and, 245–246; DEIJ environments and, 138; ethical leadership and, 188–189; lab policy and, 138–139; sexual harassment, 186, 189

# INDEX

Hartman, Laura Pincus, 181
Harvard Negotiation Project, 174
Haughton, Dan, 230, 235
Hay/McBer, 42
health, prioritizing, 23–26
Heen, Sheila, 106–107, 174–175
hidden curriculum, 139–140, 142, 200
Hoyt, Crystal, 231
Hutchins, Holly, 233
hypervisibility, 137

ideal workplace, 6–7
identity: DEIJ environments and, 129–131, 134, 136–137, 141; double jeopardy and, 235–236; empowerment and, 258–259; feedback and, 93–94; intersectionality and, 130–131, 236, 243, 259; marginalized identities, 235–236; privilege and, 235–237, 241
impostor syndrome, 86–89, 90, 233
inclusion. *See* diversity, equity, inclusion, and justice
integrity, 182, 186, 188, 193–194, 199
intersectionality, 130–131, 236, 243, 259
intrinsic motivation, 67–69, 71, 72–74, 94
introverts, 31, 40, 259

Jack, Anthony Abraham, 131
Jacobson, Lenore, 147, 220
Jeffries-EL, Malika, 77–78
job crafting, 20
Jordan, Michael, 91–92
journal editor, serving as, 19–20, 23, 205
juggling tasks, 22–23
justice. *See* diversity, equity, inclusion, and justice

Keyes, Ralph, 229–230
kindness: in DEIJ environments, 127–128, 136, 137; when giving feedback, 150–153, 157
Küüsmaa-Schildt, Maria, 26

lab culture, 122–123; aligning policies with, 264; building and improving, 99, 117, 141–142, 206, 239; shared leadership and, 119, 122–123
lab policy manuals, 121–122, 125, 264, 268, 270; communication and, 202, 208, 210–211; ethical leadership and, 191, 196; feedback and, 142, 158
Lahom, Tamra (Blue), 134
Lau, Alice Man Sze, 146
leadership strengths, 29–31; action items, 45–46; assessment tools for,

leadership strengths (*continued*) 33–42, 45; Big Five personality test and, 35–36, 37; Birkman Method and, 39–42; CliftonStrengths (StrengthsFinder) test and, 37–39; Enneagram framework and, 36–37; introversion and extroversion, 31, 35, 40, 259; weaknesses and, 32–33, 182; "weirdness" and, 29–31, 41–42, 45, 46

leadership styles, 42–45; affiliative leadership, 43, 44; authoritative leadership, 43, 44; coaching leadership, 43, 44; coercive leadership, 43, 44; democratic leadership, 43, 44; pacesetting leadership, 43, 44. *See also* leadership strengths

leadership training. *See* mentoring

leading up, 7, 204–205, 249

Lencioni, Patrick, 122–123

lifelong learning, 6, 280–281

Livingston, Robert, 235–236

Loersch, Chris, 55

mastery, 69–71, 74, 75, 94, 95

McMillan, Ron, 167–168, 170

mental health, 3; burnout, 68, 73–74, 233; communication and, 196, 206–207; confidentiality and, 246; crisis coaching example, 252–253; DEIJ environments and, 140–141, 142; ethical leadership and, 191–192, 196; feedback and, 153, 155; personality disorders, 175–177; prioritizing, 23–26; role models and, 231; time management and, 21, 22–26, 27; vulnerability and, 238

mentoring, 217–219; action items, 227–228; challenging situations and, 243, 244, 248–253; checking in with mentors, 281–282; commitment and, 272–282; DEIJ environments and, 131, 133, 134, 140, 142; ethical leadership and, 181, 194; feedback and, 94–95, 97–102, 104, 106; goal setting and, 50–51, 53, 57; graduated learning and, 265–268; mindset and, 219–220, 227; outside experts and resources, 225–227; ownership and, 122, 124, 125; role models and, 221, 227, 275–277; teaching leadership skills, 222–225, 228; time management and, 20; vulnerability and, 233, 240, 241

mindset, 6, 80–81; communication and, 204; conflict resolution and, 167; feedback and, 93–94, 146–148, 153; mentoring and, 219–220, 227;

resilience and, 80–82; vulnerability and, 231
Moore, Jeff, 272–274, 278–279
morality. *See* ethical leadership
moral manager, 181
moral person, 181
Morton, Terrell, 136–137
motivated blindness, 182
motivation, 61–62; action items, 74–75; autonomy and, 69–71, 74, 75; burnout and, 68, 73–74; creativity and, 63, 66, 69, 71, 72; extrinsic motivation, 67–69, 72–73; intrinsic motivation, 67–69, 71, 72–74, 94; leading versus lagging measures and, 73; mastery and, 69–71, 74, 75; procrastination and, 24, 62–64, 67, 74–75; purpose and, 69–71, 74, 75; the "why" of, 64–67, 74–75
motivational contagion, 72
mutual purpose, 168–170, 172–178, 200, 250

narcissistic personality disorder (NPD), 175–177
negativity bias, 85

obsessive-compulsive personality disorder (OCPD), 175–176
O'Driscoll, Michael, 155

Ones, Deniz, 31
organizational psychology, 64, 187–188

pacesetting leadership, 43, 44
Packard, Becky Wai-Ling, 133
Patterson, Kerry, 167–168, 170
personality and strengths tests, 33–42, 45; Big Five, 35–36, 37; Birkman Method, 39–42; Clifton-Strengths (StrengthsFinder), 37–39; Enneagram framework, 36–37
personality disorders, 175–177
physical safety, 131
Pink, Dan, 69
policy manuals. *See* lab policy manuals
power dynamics and imbalances, 13, 20–21, 104, 162
priority grid exercise, 17–20
procrastination, 24, 62–64, 67, 74–75
psychological safety. *See* safety
publication goals and, 73, 114, 118–119, 125
purpose: feedback and, 94, 95; motivation and, 69–71, 74, 75; mutual purpose, 168–170, 172–178, 200, 250

Pychyl, Tim, 63
Pygmalion effect, 147, 220

research ethics, 185, 186–187, 190, 193, 199, 208
resilience, 76–77; action items, 89–90; imposter syndrome and, 86–89, 90; mindset and, 80–82. *See also* failure; success
retreats: group retreats, 30–31, 118–122, 143–145, 151, 207–208; writing retreats, 61
role models: commitment and, 275–276; mentors versus, 275–277; resilience and, 259; vulnerability and, 231, 237. *See also* mentoring
Roosevelt, Teddy, 56
Rosenthal, Robert, 147, 220
Rosette, Ashleigh Shelby, 235–236
Ryan, Richard, 67, 69

Sadler, Royce, 145
safety: communication and, 209–210, 212–213; double jeopardy and, 236; feedback and, 103–105; lab safety, 149, 186, 190–191, 207; psychological safety, 103–105, 115–116, 137, 160, 229; time scale and, 115–116; trust and, 115–116

"sage on the stage" model, 114
Scott, Kim, 151–152
shared leadership, 113–114, 117–120; action items, 125–126; lab culture and, 119, 122–123; lab policy manuals and, 121–122, 125; psychological safety and, 115–116; publication goals and, 114, 118–119, 125; research direction and, 118; trust and, 115–117
Shaw, George Bernard, 198
Situation-Behavior-Impact (SBI), 154, 156, 158, 247, 251, 282
SMART goals, 52–53, 59, 99
Stafford, Laura, 198
stakeholders, 17–18
Stanier, Michael Bungay, 16
Stone, Douglas, 106–107
strengths, leadership. *See* leadership strengths; leadership styles
StrengthsFinder (CliftonStrengths), 37–39
styles, leadership. *See* leadership strengths; leadership styles
success, 84–85; enjoying and sharing, 85–86; imposter syndrome and, 86–89
Switzler, Al, 167–168, 170

task conflict, 160
templates: email "no" templates, 28; presentation slide templates, 268; proposal templates, 265
temporal discounting, 13–15
Tenbrunsel, Ann, 182
tenure process, 272–275
time management, 11–12; action items, 28; challenges of, 12; email and, 22; every yes is a no, 15–16; future self-continuity and, 12–15; juggling metaphor, 22–23; mental health and, 21, 22–26, 27; organization systems, 26–27; priority grid exercise, 17–20; prioritizing health, 23–26; response formula for saying no, 20–21; temporal discounting and, 13–15; unexpected requests and, 22–23
Title IX, 139, 189, 250–251
Treviño, Linda Klebe, 181, 221

utilitarianism, 184–185

vision, 6, 259, 274; authoritative leadership and, 43; goal setting and, 51; resilience and, 76, 79
Voss, Chris, 169–170
vulnerability, 229–231; action items, 240–241; attainability, 231–232; choosing vulnerability, 231–239; privilege and identity, 235–237, 241; sharing struggles, 233–235; unchosen vulnerability, 239–240

Wallace, Amy, 150–151
well-being, 132, 140, 142, 191, 207, 231
*WorkLife* (podcast), 29, 174
Wrzesniewski, Amy, 20

Young, Valerie, 87

Zenger, John, 32, 182
zero-sum games, 58, 282